Lecture Notes in Biomathematics

Managing Editor: S. Levin

85

Shripad Tuljapurkar

Population Dynamics in Variable Environments

Springer-Verlag
New York Berlin Heidelberg London Paris Tokyo Hong Kong

Author
Shripad Tuljapurkar
Biological Sciences
Stanford University
Stanford, CA 94305, USA

and
Graduate Group in Demography
University of California
Berkeley, CA 94720, USA

Mathematics Subject Classification (1980): 60G35, 60J70; 15A52

ISBN 3-540-52482-7 Springer-Verlag Berlin Heidelberg New York Tokyo
ISBN 0-387-52482-7 Springer-Verlag New York Heidelberg Berlin Tokyo

Printing and binding: Beltz Offsetdruck, Hemsbach/Bergstr.
2146/3140-543210 – Printed on acid-free paper

Contents

List of Figures

List of Tables

1

INTRODUCTION

Demography relates observable facts about individuals to the dynamics of populations. If the dynamics are linear and do not change over time, the classical theory of Lotka (1907) and Leslie (1945) is the central tool of demography. This book addresses the situation when the assumption of constancy is dropped. In many practical situations, a population will display unpredictable variation over time in its vital rates, which must then be described in statistical terms. Most of this book is concerned with the theory of populations which are subject to random temporal changes in their vital rates, although other kinds of variation (*e.g.*, cyclical) are also dealt with. The central questions are: how does temporal variation work its way into a population's future, and how does it affect our interpretation of a population's past.

The results here are directed at demographers of humans and at population biologists. The uneven mathematical level is dictated by the material, but the book should be accessible to readers interested in population theory. (Readers looking for background or prerequisites will find much of it in Hal Caswell's *Matrix population models: construction, analysis, and interpretation* (Sinauer 1989)). This book is in essence a progress report and is deliberately brief; I hope that it is not mystifying. I have not attempted to be complete about either the history or the subject, although most significant results and methods are presented. I have tried to be consistent in notation, but there are symbols which mean different things in different places. They should be clearly defined.

The table of contents is detailed and should serve as a guide to the reader. Here I comment on features not obvious from that table. Chapter 2 contains background material on classical demography in a form different from standard treatments, especially the material on convergence. Chapter 3 discusses reproductive value in deterministically varying environments; the material on cycles deals only with the use of perturbation methods. Chapter 4 is a rather abstract summary of the formal stochastic demography, which is fleshed out in later chapters. Chapter 5 mentions interesting empirical studies on many organisms. The results on populations which reproduce by cloning in Chapter 6 are new and there is great scope for developing the theory. Chapter 7 ends with a result I find tantalizing and which I hope someone will be able to use to compute stochastic growth rate. The detailed special cases in Chapter 8 are fascinating; note especially the singularity of distributions and the singularity of the limit where ergodicity is lost. The growth rate formula in Chapter 9 should be interesting for practical estimation. The example in Chapter 10, Section 2, is fun; I would like to hear about other surprises like that. The results in Chapter 11, Section 2, are new and should be at least as useful as the classical sensitivity results. Most of the material in Chapter 13 has been little explored and has lots of applications. I had hoped to write another chapter on extinction probabilities but this book has to end somewhere.

ACKNOWLEDGEMENT

I am in the debt of many people and organizations. For the book, Carl Boe worked hard on all aspects, made the production possible, and tried to infect me with his standards of perfection; Cheryl Nakashima did most of the typing in her inimitably skilled way; A. Meredith John read carefully through most of the text, corrected many errors, and attempted to keep me on the stylistic straight-and-narrow. Grants from the National Institutes of Health (NICHHD 16640 and 00639) provided most of the time and resources. Marc Feldman at Stanford and the Berkeley demographers - Gene Hammel, Ron Lee and Ken Wachter - have been extremely generous with their hospitality and support. Shubha Tuljapurkar is my constant source of love and optimism. Anjali Tuljapurkar pressured me to finish, continually asking when this would be a "book" instead of piles of paper. Countless fishing trips with my buddy Bill Green have been essential medicine for my spirits.

Ray Sommerfeldt at Portland State University provided essential encouragement when I first started work in this area. Steve Orzack has argued fiercely and worked closely with me on some of this material, greatly contributing to the enjoyment of discovery. Many other colleagues have been encouraging and interested at critical times, and I am especially grateful to Hal Caswell, Ansley Coale, Joel Cohen, Jim Crow, Lev Ginzburg, Conrad Istock, Sam Karlin, Dick Lewontin, Simon Levin, Norm Slade, Pavel Smejtek, Burt Singer and Michael Turelli.

This book is dedicated to my mother, Rajeshwari Tuljapurkar, and to the memory of my father, Digambar Vinayak Tuljapurkar.

2

BEGINNINGS: CLASSICAL THEORY

Most of this book deals with demography in discrete time, *i.e.*, time is divided into successive intervals of some equal length τ, and the times τ, 2τ, etc. are labeled as discrete times 1, 2, etc. Demographers usually drop the τ for convenience, as will I, but note that this practice obscures the dimensions (*i.e.*, units) of all the objects in our equations. This chapter summarizes demographic theory in discrete time with *constant* rates, emphasizing terminology and results which I later use freely without further explanation. For details consult Keyfitz (1968) or the more compact account of Pollard (1973). Coale's (1972) book uses continuous time but is valuable.

1 General Discrete Time Model

Consider a population divided into k discrete classes, which may be age, size or similar groups. At (discrete) time t let $n_t(i)$, $i = 1, \ldots, k$, be the numbers of individuals in class i, and form the vector $\boldsymbol{n}_t$. Assume that the transition rate between class j at time t and class i at time $t+1$ is a (time-independent) number $b_{ij} \geq 0$ so that

$$n_{t+1}(i) = \sum_j b_{ij} n_t(j). \tag{2.1.1}$$

Using the matrix $\boldsymbol{b} = (b_{ij})$ this becomes

$$\boldsymbol{n}_{t+1} = \boldsymbol{b}\boldsymbol{n}_t. \tag{2.1.2}$$

This equation applies only to populations satisfying various assumptions, *e.g.*, we count only one sex in a population closed to immigration and emigration, ignoring density effects and differences between individuals. I will often call matrices such as $\boldsymbol{b}$ projection matrices.

Given an initial population vector $\boldsymbol{n}_0$, the solution of (2.1.2) is

$$\boldsymbol{n}_t = \boldsymbol{b}^t \boldsymbol{n}_0. \tag{2.1.3}$$

In addition to the vector $\boldsymbol{n}_t$ we are interested in total population size

$$m_t = (\boldsymbol{e}, \boldsymbol{n}_t) = \sum_i n_t(i). \tag{2.1.4}$$

Here I use the **scalar product** of vectors $\boldsymbol{x}$, $\boldsymbol{y}$ defined as

$$(\boldsymbol{x}, \boldsymbol{y}) = \sum_i x^*(i)y(i), \tag{2.1.5}$$

where the star superscript denotes a **complex conjugate**, and $\boldsymbol{e}$ is a vector of 1s,

$$\boldsymbol{e} = \begin{pmatrix} 1 \\ 1 \\ \vdots \\ 1 \end{pmatrix} = (1, 1, \ldots, 1)^T. \tag{2.1.6}$$

As in (2.1.6), vectors here are column vectors, and the superscript T means "transpose". A dagger (†) superscript will indicate a complex conjugate transpose. We are also interested in the population **structure**, defined as

$$\boldsymbol{y}_t = \boldsymbol{n}_t / (\boldsymbol{e}, \boldsymbol{n}_t) = \boldsymbol{n}_t / m_t. \tag{2.1.7}$$

2 Dynamics and Spectral Decomposition

The properties of $\boldsymbol{n}_t$, $\boldsymbol{y}_t$ and m_t depend on $\boldsymbol{b}$. A full mathematical account of such nonnegative matrices is given by Seneta (1981), briefer ones by Karlin and Taylor (1975, Appendix) and Lancaster (1969). I focus on two cases. The first is when $\boldsymbol{b}$ is **primitive**, *i.e.*, there is some integer $g \geq 1$ such that every element of $\boldsymbol{b}^g$ is positive. In this case the Perron-Frobenius theorem (see references above) states that there is a positive number λ_0 and vectors $\boldsymbol{u}_0$, $\boldsymbol{v}_0$ with all components positive, such that

$$\boldsymbol{b}\boldsymbol{u}_0 = \lambda_0 \boldsymbol{u}_0, \quad \boldsymbol{v}_0^T \boldsymbol{b} = \lambda_0 \boldsymbol{v}_0^T. \tag{2.2.1}$$

Call λ_0 the **dominant eigenvalue** of $\boldsymbol{b}$; it exceeds in magnitude all other eigenvalues of $\boldsymbol{a}$, which are, of course, **subdominant eigenvalues.** It is convenient to normalize the vectors in (2.2.1) so that

$$(\boldsymbol{e}, \boldsymbol{u}_0) = 1, \tag{2.2.2}$$

and

$$(\boldsymbol{v}_0, \boldsymbol{u}_0) = 1. \tag{2.2.3}$$

For primitive $\boldsymbol{b}$ it follows that as $t \to \infty$,

$$\lambda_0^{-t} \boldsymbol{n}_t \to (\boldsymbol{v}_0, \boldsymbol{n}_0)\boldsymbol{u}_0, \tag{2.2.4}$$

$$\boldsymbol{y}_t \to \boldsymbol{u}_0, \tag{2.2.5}$$

$$\lambda_0^{-t} m_t \to (\boldsymbol{v}, \boldsymbol{n}_0) = (\boldsymbol{v}_0, \boldsymbol{y}_0) m_0. \tag{2.2.6}$$

The role played by $\boldsymbol{v}_0$ in (2.2.6), of weighting the contribution of initial population structure to eventual size, defines $\boldsymbol{v}_0$ as the **reproductive value** vector. Clearly u_0 is the **stable population structure**. The long run geometric growth rate of population is λ_0; the long run logarithmic growth rate of population is

$$\lim_{t\to\infty} \frac{1}{t} \log(m_t/m_0) = \log \lambda_0 \tag{2.2.7}$$

$$= r_0, \tag{2.2.8}$$

where we use the ubiquitous notation, for the Malthusian parameter in (2.2.8).

We learn more about the primitive case from the **spectrum** of b, *i.e.*, the set of its eigenvalues which can be ordered as

$$\lambda_0 > |\lambda_1| \geq |\lambda_2| \geq \cdots \geq |\lambda_{k-1}|. \tag{2.2.9}$$

There is a **simple spectral decomposition** of b (Lancaster (1969), Pollard (1973), Keyfitz (1968),

$$b = \lambda_0 u_0 v_0^T + \lambda_0 q. \tag{2.2.10}$$

Here

$$q^t \to 0 \quad \text{as} \quad t \to \infty, \tag{2.2.11}$$

with the rate of decrease given by

$$q^t \sim 0(|\lambda_1|/\lambda_0)^t \quad \text{as} \quad t \to \infty. \tag{2.2.12}$$

This tells us that $(|\lambda_1|/\lambda_0)$ is the rate at which y_t approaches u_0. If we let

$$\lambda_1 = e^{r_1 + i\omega_1}, \tag{2.2.13}$$

then

$$\sum_i |y_t(i) - u_0(i)| \sim e^{-(r_0 - r_1)t} e^{i\omega_1 t} \tag{2.2.14}$$

as $t \to \infty$. I call $(r_0 - r_1)$ (or, sometimes, $|\lambda_1|/\lambda_0$) the classical **damping rate of transient** population oscillations, while ω_1 is their **angular frequency** and $(2\pi/\omega_1)$ is their **period**. Coale (1972) and Wachter (1989) have studied properties of r_1 and ω_1.

Yet more detail about the primitive case can be obtained from a **full spectral decomposition** of b, which is possible when *all* the eigenvalues are distinct. (This is actually typical of a matrix; a matrix which has repeated (degenerate) eigenvalues can usually be made into one with distinct eigenvalues by adding very small perturbations to its elements.) In this case every eigenvalue λ_i has corresponding left, right eigenvectors v_i, u_i, and we normalize them to ensure

$$(e, u_i) = (v_i, u_i) = 1, \quad i = 0, 1, \ldots, k-1. \tag{2.2.15}$$

Then the full spectral decomposition is

$$b = \sum_i \lambda_i u_i v_i^\dagger = \sum_i \lambda_i s_i, \tag{2.2.16}$$

with

$$s_i = u_i v_i^\dagger = s_i^2,$$
$$s_i s_j = 0 \quad \text{when} \quad i \neq j, \tag{2.2.17}$$

and

$$\sum_i s_i = I. \tag{2.2.18}$$

The s_i are **projection matrices** which tell us how far a particular age structure differs from u_0. Here the matrix q of the simple spectral decomposition (2.2.10) can be explicitly written as

$$q = \sum_{i \geq 1} (\lambda_i/\lambda_0) s_i. \tag{2.2.19}$$

The second general class of matrices b are those which are not primitive, so that every integer power of b has some zeroes. Imprimitive matrices are further divided into reducible and irreducible ones; I will usually deal only with irreducible ones. The main feature of an imprimitive matrix b is that it will not (in general) have a single dominant eigenvalue. Instead there are, say, h eigenvalues, one of which is a real λ_0, and $\lambda_1, \lambda_2, \ldots, \lambda_{h-1}$ are complex but have magnitude λ_0. In this case we have

$$\lambda_j = \lambda_0 e^{(2\pi i j/h)}, \qquad j = 0, 1, \ldots, h-1, \tag{2.2.20}$$

and

$$|\lambda_j| < \lambda_0, \qquad j = h, h+1, \ldots, k-1. \tag{2.2.21}$$

The consequence of (2.2.20) is that powers b^t have a cyclical pattern with period h as t increases. A population described by such b therefore shows persistent cycles whose period is h but whose amplitude depends on the initial state. The time average of population structure still has a "stable" behavior since

$$\lim_{t \to \infty} \frac{1}{t} \sum_{m=1}^{t} \lambda_0^{-m} b^m = u_0, \tag{2.2.22}$$

where u_0 is the right eigenvector of b corresponding to λ_0. See Cull and Vogt (1973) for a fuller discussion of such cycling behavior.

3 A Geometrical View of Convergence

The convergence of population structure $\boldsymbol{y}_t$ from any initial $\boldsymbol{y}_0$ to $\boldsymbol{u}_0$ is central to the classical theory and is usually viewed in terms of a "decreasing distance," as in equation (2.2.14). I now introduce a different picture of convergence which has more geometrical content and which generalizes to the time-dependent case. I will work with the model (2.1.2) with a primitive $\boldsymbol{b}$ having distinct eigenvalues, so that (2.2.15)–(2.2.18) hold. The right eigenvectors ($\boldsymbol{u}_\alpha$, $\alpha = 0, 1, \ldots, k-1$) are a linearly independent set, so we can expand any initial population vector as

$$\boldsymbol{n}_0 = \sum_\alpha c_{0\alpha} \boldsymbol{u}_\alpha, \tag{2.3.1}$$

with

$$c_{0\alpha} = (\boldsymbol{v}_\alpha, \boldsymbol{n}_0).$$

Over time the population vector changes to

$$\boldsymbol{n}_t = \sum_\alpha c_{t\alpha} \boldsymbol{u}_\alpha, \tag{2.3.2}$$

with

$$\begin{aligned} c_{t\alpha} &= \lambda_\alpha c_{t-1,\alpha} \\ &= \lambda_\alpha^t c_{0\alpha}. \end{aligned} \tag{2.3.3}$$

I now consider two distinct nonparallel initial vectors, say the one in (2.3.1) and

$$\boldsymbol{n}'_0 = \sum_\alpha c'_{0\alpha} \boldsymbol{u}_\alpha. \tag{2.3.4}$$

I will examine the rate of convergence at which the population structures evolving from $\boldsymbol{n}_0$ and $\boldsymbol{n}'_0$ approach each other. A systematic way of doing this is to study the area enclosed by the vectors $\boldsymbol{n}_t$ and $\boldsymbol{n}'_t = \boldsymbol{b}^t \boldsymbol{n}'_0$. A general definition of the area between vectors $\boldsymbol{x}$, $\boldsymbol{y}$ can be given in terms of their **Kronecker product**

$$\boldsymbol{x} \otimes \boldsymbol{y} = \begin{pmatrix} x_1 \boldsymbol{y} \\ x_2 \boldsymbol{y} \\ \vdots \end{pmatrix} = \begin{pmatrix} x_1 y_1 \\ x_1 y_2 \\ \vdots \\ x_1 y_k \\ x_2 y_1 \\ x_2 y_2 \\ \vdots \end{pmatrix}. \tag{2.3.5}$$

The **wedge product** of $\boldsymbol{x}$ and $\boldsymbol{y}$ is

$$\boldsymbol{x} \wedge \boldsymbol{y} = \boldsymbol{x} \otimes \boldsymbol{y} - \boldsymbol{y} \otimes \boldsymbol{x}. \tag{2.3.6}$$

As defined in (2.3.6), $(\boldsymbol{x} \wedge \boldsymbol{y})$ is a vector of k^2 components of which k equal zero. The **area** spanned by the parallelepiped of which $\boldsymbol{x}$ and $\boldsymbol{y}$ are sides is given by

$$A(\boldsymbol{x}, \boldsymbol{y}) = \frac{1}{\sqrt{2}} \|\boldsymbol{x} \wedge \boldsymbol{y}\|, \tag{2.3.7}$$

where the (Euclidean) norm of a vector $\boldsymbol{x}$ is used:

$$\|\boldsymbol{x}\| = (\boldsymbol{x}, \boldsymbol{x})^{\frac{1}{2}}. \tag{2.3.8}$$

Returning to the population vectors $\boldsymbol{n}_t$ and $\boldsymbol{n}'_t$, consider

$$\begin{aligned} \boldsymbol{n}_t \wedge \boldsymbol{n}'_t &= \left(\sum_{\alpha} \lambda_{\alpha}^t c_{0\alpha} \boldsymbol{u}_{\alpha} \right) \wedge \left(\sum_{\beta} \lambda_{\beta}^t c'_{0\beta} \boldsymbol{u}_{\beta} \right) \\ &= \sum_{\alpha \neq \beta} \lambda_{\alpha}^t \lambda_{\beta}^t c_{0\alpha} c'_{0\beta} (\boldsymbol{u}_{\alpha} \wedge \boldsymbol{u}_{\beta}). \end{aligned} \tag{2.3.9}$$

Here the restriction on the last sum leaves out all zero terms (since $\boldsymbol{x} \wedge \boldsymbol{x} = 0$). Now the terms on the right-side of (2.3.9) change with time at different exponential rates, the fastest-changing terms being those with $\alpha = 0$, $\beta = 1$ and $\alpha = 1$, $\beta = 0$. Letting population sizes for the two different population vectors be m_t, m'_t, consider

$$\boldsymbol{y}_t \wedge \boldsymbol{y}'_t = (\boldsymbol{n}_t / m_t) \wedge (\boldsymbol{n}'_t / m'_t). \tag{2.3.10}$$

From (2.3.9) we get for large t,

$$(\boldsymbol{y}_t \wedge \boldsymbol{y}'_t) \sim (\lambda_1/\lambda_0)^t (c_{00} c'_{01} - c_{01} c'_{00})(\boldsymbol{u}_0 \wedge \boldsymbol{u}_1) + c.c., \tag{2.3.11}$$

where "*c.c.*" stands for "complex conjugate of preceding term." Since $\boldsymbol{n}_0$ and $\boldsymbol{n}'_0$ are nonparallel, the area between them is nonzero and thus the right-side of (2.3.11) is nonzero. Asymptotically, the area between the structures $\boldsymbol{y}_t$ and $\boldsymbol{y}'_t$ goes to zero at a rate

$$\begin{aligned} \lim_{t \to \infty} \frac{1}{t} \log A(\boldsymbol{y}_t, \boldsymbol{y}'_t) &= \lim_{t \to \infty} \frac{1}{t} \log \|\boldsymbol{y}_t \wedge \boldsymbol{y}'_t\| \\ &= -(\log \lambda_0 - \log |\lambda_1|) && (2.3.12) \\ &= -(r_0 - r_1). && (2.3.13) \end{aligned}$$

The reader may show by extension of this argument that the volume enclosed by a triplet of three initially distinct nonparallel population structures goes to zero at a rate

$$\lim_{t \to \infty} \log \|\boldsymbol{y}_t \wedge \boldsymbol{y}'_t \wedge \boldsymbol{y}''_t\| = -(r_0 - r_1) - (r_0 - r_2), \tag{2.3.14}$$

where $\boldsymbol{y}_t$, $\boldsymbol{y}'_t$, $\boldsymbol{y}''_t$ are the structure vectors, $r_2 = \log |\lambda_2|$, and the wedge product is defined by extension [*i.e.*, $\boldsymbol{x} \wedge \boldsymbol{y} \wedge \boldsymbol{z} = \boldsymbol{x} \wedge (\boldsymbol{y} \wedge \boldsymbol{z})$]. The argument can be extended to cover hypervolumes enclosed by k–triplets of vectors. This approach to convergence will be useful in the case of random rates.

4 Leslie Matrices and Age Structure

Nothing in Sections 2.1–2.3 is restricted to age structured populations. If the population is age structured, then $\boldsymbol{b}$ is the usual **Leslie matrix,**

$$\boldsymbol{b} = \begin{pmatrix} f_1 & f_2 & \cdots & f_k \\ p_1 & 0 & \cdots & 0 \\ 0 & \cdot & p_{k-1} & 0 \end{pmatrix}. \tag{2.4.1}$$

I will *always assume* that only reproductive and prereproductive ages are being counted, *i.e.*, $f_k > 0$. The subdiagonal elements in (2.4.1) are survival rates with $0 < p_i \leq 1$ for all i. A sufficient condition for primitive $\boldsymbol{b}$ is that two consecutive f_i are nonzero. A general expression for g, the smallest integer such that $\boldsymbol{b}^g > 0$, is given by Hansen (1983) following Tuljapurkar (1981).

It is convenient to recall here some well-known formulae. Define **survivorship,** *i.e.*, chance of survival to age class i, by

$$\ell_i = \begin{cases} 1, & i = 1, \\ p_1 p_2 \ldots p_{i-1}, & i \geq 2. \end{cases} \tag{2.4.2}$$

The **net maternity** value at age i is defined as

$$\phi_i = \ell_i m_i. \tag{2.4.3}$$

For Leslie matrices, λ_0 is the largest real solution of the **characteristic equation**

$$\sum_{i=1}^{k} \lambda^{-i} \phi_i = 1. \tag{2.4.4}$$

The other solutions of (2.4.4) are the subdominant eigenvalues λ_i. For each $\alpha = 0, 1, \ldots, k-1$, the eigenvectors $\boldsymbol{u}_\alpha$, $\boldsymbol{v}_\alpha$ are given explicitly by defining for $i = 1, \ldots, k$,

$$\widehat{u}_\alpha(i) = \lambda_\alpha^{-i+1} \ell_\alpha, \tag{2.4.5}$$

$$u_\alpha(i) = \widehat{u}_\alpha(i) \Big/ \Big(\sum_j \widehat{u}_\alpha(j) \Big), \tag{2.4.6}$$

$$\widehat{v}_\alpha(i) = (1/\ell_i) \Big(\sum_{j=i}^{k} f_j \ell_j \lambda_\alpha^{i-j-1} \Big), \tag{2.4.7}$$

$$T_\alpha = \sum_i \widehat{u}_\alpha(i) \widehat{v}_\alpha(i), \tag{2.4.8}$$

$$v_\alpha(i) = \widehat{v}_\alpha(i) \Big(\sum_j u_\alpha(j) / T_\alpha \Big). \tag{2.4.9}$$

This ensures that $(\boldsymbol{e}, \boldsymbol{u}_\alpha) = (\boldsymbol{v}_\alpha, \boldsymbol{u}_\alpha) = 1$ for all α. The quantity T_0 (put $\alpha = 0$ in (2.4.8)) is the **mean length of generation**

$$T_0 = \sum_{i=1}^{k} i \lambda_0^{-i} \phi_i. \tag{2.4.10}$$

3

DETERMINISTIC TEMPORAL VARIATION

I now consider what happens when the fixed matrix $\boldsymbol{b}$ of (2.1.2) is replaced by a deterministically varying sequence of matrices. This is a long-standing problem with Norton (1928), Coale (1957), and Lopez (1961) being the classical contributions. Golubitsky *et al.* (1977), Hajnal (1976), Kim and Sykes (1976), Seneta (1981), Cohen (1979b), Tuljapurkar (1984) and Kim (1987) are more recent explorations. This work is a prerequisite to the study of random rates. I first consider general variation, and then cyclical variation.

1 General Temporal Changes

The time-varying version of the population model is

$$\boldsymbol{n}_{t+1} = \boldsymbol{b}_{t+1}\boldsymbol{n}_t. \tag{3.1.1}$$

Here I use $(t+1)$ as an index for $\boldsymbol{b}$ because $\boldsymbol{b}_{t+1}$ contains vital rates which act on $\boldsymbol{n}_t$ in the interval $[t, t+1)$ to produce $\boldsymbol{n}_{t+1}$. Starting with some initial $\boldsymbol{n}_0$, suppose that we have a particular sequence of matrices $\boldsymbol{b}_1, \boldsymbol{b}_2, \ldots, \boldsymbol{b}_t$. Successive age structures obey the equation

$$\boldsymbol{y}_{t+1} = \boldsymbol{b}_{t+1}\boldsymbol{y}_t / (\boldsymbol{e}, \boldsymbol{b}_{t+1}\boldsymbol{y}_t), \tag{3.1.2}$$

and growth rates are

$$\eta_{t+1} = m_{t+1}/m_t = (\boldsymbol{e}, \boldsymbol{b}_{t+1}\boldsymbol{y}_t), \tag{3.1.3}$$

where m_t, as before, is total population at time t.

The key to making sense out of this time-varying situation is to ask *when* the demographic process *forgets* its initial state. The answer is, when the sequence of matrices in (3.1.1) obeys **demographic weak ergodicity.** This happens if the product matrix $\boldsymbol{b}_t\boldsymbol{b}_{t-1}\ldots\boldsymbol{b}_1$ (which determines $\boldsymbol{n}_t$) ends up having all entries positive for large t. In that case, there is a stable but time-varying age structure sequence $\widehat{\boldsymbol{y}}_t$ such that $\boldsymbol{y}_t \rightarrow \widehat{\boldsymbol{y}}_t$ in (3.1.2), independent of $\boldsymbol{y}_0$, for large t.

There are various sufficient conditions for demographic weak ergodicity, reviewed by Seneta (1981). In this book, I use Hajnal's (1976) notion of an **ergodic set**: this is a collection of matrices accompanied by an integer

g, such that every product of g matrices from the collection is a matrix with every element positive. Thus, demographic weak ergodicity in (3.1.1) is assured if all the matrices which can appear belong to an ergodic set. The simplest example (and a useful one) of an ergodic set is a collection of nonnegative matrices which all have positive elements and zeroes in the same locations, with any one matrix being primitive. If we construct a matrix with 1s where this collection has positive entries and 0's whenever this collection has zeroes, we obtain what is called the **incidence matrix** for the collection. A nonnegative matrix is primitive if and only if its incidence matrix is primitive, so the ergodicity of our collection is established.

Given demographic weak ergodicity, it is possible and useful to extend the concept of reproductive value to the time-varying case. Recall that the vital rates have the interpretation

$$(\boldsymbol{b}_t)_{ij} = \text{Number of class } i \text{ individuals at time } t \text{ per class } j \text{ individual at time } t-1. \tag{3.1.4}$$

Consider now a quantity $w_t(i)$, to be called the undiscounted reproductive value of an individual in class i at time t, and defined as the total number of descendants (*i.e.* children, grandchildren, and so on) produced by an individual who is in class i at time t. From this definition and the interpretation of the vital rates above, it follows that

$$w_t(i) = \sum_j (b_{t+1})_{ji} w_{t+1}(j). \tag{3.1.5}$$

The vector form of this recursion is

$$\boldsymbol{w}_t = \boldsymbol{b}_{t+1}^T \boldsymbol{w}_{t+1} = \boldsymbol{b}_{t+1}^T \boldsymbol{b}_{t+2}^T \ldots \boldsymbol{b}_{t+m}^T \boldsymbol{w}_{t+m}. \tag{3.1.6}$$

To deal with the obvious possibility that the $\boldsymbol{w}$'s are likely to be unbounded, define instead the normalized (or discounted) reproductive value vector

$$\boldsymbol{v}_t = \boldsymbol{w}_t / (\boldsymbol{e}, \boldsymbol{w}_t) \tag{3.1.7}$$

which follows the recursion

$$\boldsymbol{v}_t = \boldsymbol{b}_{t+1}^T \boldsymbol{v}_{t+1} / (\boldsymbol{e}, \boldsymbol{b}_{t+1}^T \boldsymbol{v}_{t+1}). \tag{3.1.8}$$

Take two times, k and $\ell = (k+m) > k$, and consider

$$\boldsymbol{z}(k,\ell) = \boldsymbol{b}_\ell \boldsymbol{b}_{\ell-1} \ldots \boldsymbol{b}_{k+1} \tag{3.1.9}$$

which is a product of m matrices. Demographic weak ergodicity implies (Hajnal 1976) that this product will have its rows all proportional as m increases. Thus there is a number $\rho(k,\ell)$ and vectors $\boldsymbol{v}(k,m)$ and $\boldsymbol{u}(\ell,m)$ such that

$$\boldsymbol{z}(k,\ell) \sim \rho(k,\ell) \boldsymbol{u}(\ell,m) \boldsymbol{v}^T(k,m) \quad \text{as} \quad m \uparrow. \tag{3.1.10}$$

Further we know there is stability of age structure, meaning that $\boldsymbol{u}(\ell, m)$ approaches some $\boldsymbol{u}(\ell)$ asymptotically independent of m; similarly $\boldsymbol{v}(k, m)$ approaches some $\boldsymbol{v}(k)$. Thus asymptotically for large m

$$\boldsymbol{z}(k, \ell) \sim \rho(k, \ell)\boldsymbol{u}(\ell)\boldsymbol{v}^T(k). \tag{3.1.11}$$

Numerical insights into (3.1.11) are to be found in Kim and Sykes (1976). Suppose now that we start with a population vector $\boldsymbol{n}^*$ at time $t = k$. Then at time ℓ we have asymptotically a population vector

$$\boldsymbol{n}_\ell \sim \rho(k, \ell)\left(\boldsymbol{v}(k), \boldsymbol{n}^*\right)\boldsymbol{u}(\ell),$$

with population structure $\boldsymbol{y}_\ell \sim \boldsymbol{u}(\ell)$. The growth rate here is contained in $\rho(k, \ell)$, so that $\log \rho(k, \ell)/(\ell - k)$ is the long run growth rate for $(\ell - k) \to \infty$. The normalized reproductive value at time k is $\boldsymbol{v}(k)$. To find $\boldsymbol{v}(k)$ simply start with an arbitrary nonnegative vector $\widehat{\boldsymbol{v}} \neq 0$ at time $t = \ell \gg k$ in (3.1.8) and iterate backwards.

Notice that the vital rate matrices in (3.1.2) act to propagate population vectors forward in time, and that the stable age structure at each time is an accumulation of the past. In contrast, the transposed vital rate matrices in (3.1.8) act to propagate reproductive value backward in time, and the reproductive value at each time is a summation of the future.

2 Cyclically Changing Vital Rates

Seasonal variation will often drive marked periodic variation in a population's vital rates. Human populations can be influenced by longer economic cycles; both human and natural populations can be affected by long period climate cycles. Formal analysis of such cycles was (probably) initiated by MacArthur (1968) who studied a model with 2 age classes. Coale (1972) allowed fertility to vary cyclically and used Fourier methods to explore the dynamics. Here I summarize a discrete time extension of Coale's results due to Tuljapurkar (1985). The analysis uses a perturbation technique to develop a systematic understanding of dynamics with cyclical rates.

The model for cyclical rates is

$$\boldsymbol{n}_{t+1} = \left[\boldsymbol{b} + \boldsymbol{d}\cos(\omega t)\right]\boldsymbol{n}_t. \tag{3.2.1}$$

Here $\omega = (2\pi/T)$ is the angular frequency corresponding to the cycles of period T. The vital rates have average values contained in matrix $\boldsymbol{b}$, and their cyclical amplitudes are contained in $\boldsymbol{d}$. Since the total matrix on the right of (3.2.1) is nonnegative, one has $|d_{ij}| \leq b_{ij}$ whenever $b_{ij} \neq 0$, and $d_{ij} = 0$ otherwise. I assume that

$$\max{}_{i,j}\left(|d_{ij}|/b_{ij}\right) = g < 1, \tag{3.2.2}$$

(where we consider only indices such that $b_{ij} > 0$). If we write

$$\boldsymbol{b}_t = \boldsymbol{b} + \boldsymbol{d}\cos(\omega t), \tag{3.2.3}$$

the full solution of (3.2.1) is described by the cycle matrix,

$$\boldsymbol{c} = \boldsymbol{b}_{T-1}\boldsymbol{b}_{T-2}\ldots\boldsymbol{b}_0. \tag{3.2.4}$$

The difficulty in studying cyclical rates is precisely the complexity of $\boldsymbol{c}$; if we could find its dominant eigenvalue μ and corresponding eigenvector $\boldsymbol{u}$, we'd know the asymptotic properties of $\boldsymbol{n}_t$.

The Fourier analysis proceeds by arguing that a cycle in the vital rates of frequency ω will produce cycles in the population vector at all the *harmonic* frequencies, $(\omega, 2\omega, 3\omega, \ldots)$. Since these are discrete time cycles, their shortest period is 2 and their highest frequency is $(2\pi/2) = \pi$. Hence the highest possible harmonic M satisfies $M\omega \leq \pi$, making M the largest integer $\leq T/2$. Asymptotically, the population vector $\boldsymbol{n}_t$ has an overall growth rate λ, say, and consists of components whose frequencies are $0, \omega, 2\omega, \ldots, M\omega$:

$$\boldsymbol{n}_t \sim \lambda^t \left[\boldsymbol{h}_0 + \sum_{m=1}^{M} \boldsymbol{h}_m z^{mt} + c.c.\right], \tag{3.2.5}$$

where $z = e^{i\omega}$ and $c.c.$ (as before) is the complex conjugate of the immediately preceding term. Insert (3.2.5) into (3.2.1), use (3.2.3), and get the system of equations

$$\lambda z^m \boldsymbol{h}_m = \boldsymbol{b}\boldsymbol{h}_m + \frac{1}{2}\boldsymbol{d}\boldsymbol{h}_{m-1} + \frac{1}{2}\boldsymbol{d}\boldsymbol{h}_{m+1}, \qquad m = 0, 1, \ldots, M, \tag{3.2.6}$$

with the conventions

$$\boldsymbol{h}_{-1} = \boldsymbol{h}_1^*, \qquad \boldsymbol{h}_{M+1} \equiv 0. \tag{3.2.7}$$

This system is closed by recalling that the average projection matrix $\boldsymbol{b}$ has dominant eigenvalue λ_0 with corresponding left eigenvector $\boldsymbol{v}_0$ (see (2.2.1)–(2.2.3)). Applying $\boldsymbol{v}_0$ to the $m = 0$ version of (3.2.6) yields

$$\lambda = \lambda_0 + \frac{(\boldsymbol{v}_0, \Re e[\boldsymbol{h}_1])}{(\boldsymbol{v}_0, \boldsymbol{h}_0)}. \tag{3.2.8}$$

Here $\Re e$ indicates the real part of a complex number.

The next step is to assess the size of the harmonic components in (3.2.5). We may say that every element of $\boldsymbol{d}$ in (3.2.3) is $O(g)$, where g from (3.2.2) is the magnitude of cyclical variation in the rates. Perturbation theory (Kato 1966, see Tuljapurkar 1985 for specifics) shows that the growth rate μ and the eigenvalue $\boldsymbol{u}$ of $\boldsymbol{c}$ satisfy

$$\mu - \lambda_0^T = O(g^2),$$
$$\boldsymbol{u} - \boldsymbol{u}_0 = O(g).$$

These estimates plus (3.2.6)–(3.2.7) show that

$$\boldsymbol{h}_m = O(g^m) \qquad \text{for} \qquad m \geq 1.$$

Hence the successive harmonic components in (3.2.5) are of decreasing importance.

I focus on the main effects in (3.2.5) which are revealed by approximating

$$\boldsymbol{h}_0 \simeq \boldsymbol{u}_0, \tag{3.2.9}$$

which leads to

$$\boldsymbol{h}_1 \simeq \frac{1}{2}(\lambda_0 z - \boldsymbol{b})^{-1} \boldsymbol{d}\boldsymbol{u}_0 \tag{3.2.10}$$

and a corresponding estimate for λ from (3.2.8). Assuming that $\boldsymbol{b}$ has distinct eigenvalues, let us employ the full spectral decomposition (2.2.16) to expand the inverse matrix in (3.2.10) as

$$(\lambda_0 z - \boldsymbol{b})^{-1} = \sum_{\alpha=0}^{k-1} (\lambda_0 z - \lambda_\alpha) \boldsymbol{s}_\alpha, \tag{3.2.11}$$

so that

$$\boldsymbol{h}_1 \simeq \frac{1}{2\lambda_0} \sum_{\alpha \geq 0} \frac{1}{(z - \lambda_\alpha / \lambda_0)} \boldsymbol{s}_\alpha \boldsymbol{d}\boldsymbol{u}_0. \tag{3.2.12}$$

Since every λ_α for $\alpha > 0$ is (in general) complex, we have reduced the dominant harmonic part of $\boldsymbol{n}_t$ to a weighted sum of sinusoidal terms in (3.2.12).

The principal conclusions obtained from (3.2.12) are:

(i) From (3.2.12) we can determine the amplification and phase of population oscillations relative to the underlying vital rate cycle.

(ii) We know that every eigenvalue λ_α of $\boldsymbol{b}$ can be written as

$$\lambda_\alpha = |\lambda_\alpha| e^{i\phi_\alpha}.$$

Therefore any one of the denominators in (3.2.12) will reach its maximal value when the external cycle frequency matches the appropriate transient frequency in the population, *i.e.*, whenever we have a resonance with $\omega = \phi_\alpha$, for any α.

(iii) The major resonance occurs for the leading subdominant eigenvalue λ_1 and corresponds to a frequency $\omega \approx 2\pi / T_0$, where T_0 is the mean length of generation for an age-structured population.

(iv) The population growth rate λ from (3.2.8) is usually found to be decreased from the value of λ_0 when cyclical variation is added, *except* for cycles very near the main resonance frequency of $(2\pi / T_0)$. For frequencies less or greater than this, $\lambda < \lambda_0$ with the difference being quite substantial. For frequencies very near the main resonance frequency, cycles can produce $\lambda > \lambda_0$, the increase being moderate.

4

RANDOM RATES: GENERAL THEORY

Demographic theory with random vital rates is built on powerful general properties of random matrix products. This chapter sets out the kinds of random models I want to analyze, and summarizes general random matrix properties. Later chapters consider applications and questions which require a more concrete study of particular models.

1 Models and Questions

The classical projection model is easily generalized to include vital rates which change over time in potentially unpredictable ways. At time t, let the population vector be $\boldsymbol{N}_t$, the population structure vector be $\boldsymbol{Y}_t$ (we use capitals to show that these are random variables). Over the interval t to $t+1$ demographic processes operate on these vectors, and their overall effect is contained in a time-dependent matrix of vital rates $\boldsymbol{X}_{t+1}$. The dynamics of population are given by the equation

$$\boldsymbol{N}_{t+1} = \boldsymbol{X}_{t+1}\boldsymbol{N}_t. \tag{4.1.1}$$

The matrix subscript is $(t+1)$ rather than t to emphasize that these rates apply to the vector $\boldsymbol{N}_t$. Thus in some cases (*e.g.*, the I.I.D. model below) $\boldsymbol{N}_t$ and $\boldsymbol{X}_{t+1}$ are independent. In order to proceed, we must specify the kind of uncertainty that occurs in the vital rates. The models I consider have the following implicit structure: there is an exogenously determined time-dependent random variable called the "environment." Vital rates at each time depend on the state of the environment. Examples of subsequent interest are:

The I.I.D. Model: The entries of $\boldsymbol{X}$ are chosen randomly for each t from the same fixed (in general multivariate) distribution. There may be correlations between vital rates within each period, but there is no serial correlation between rates at different times. Here the environment is completely unpredictable. The number of possible environments can be finite (*e.g.*, a "good" and a "bad" state), or infinite (*e.g.*, if there is a continuously distributed variable like temperature).

The Markov Model: From one time interval to the next, vital rates change according to time invariant transition probabilities. There are three subcases, according as the set of possible values of vital rates is **finite**, **countable** but infinite (*e.g.*, discrete environmental states but infinitely many of them), or **uncountable** (usually continuously distributed) set. Here the environment is predictable to the extent that there is serial autocorrelation over time.

The ARMA Model: The elements of the vital rates follow a linear time series model of the ARMA type (Box and Jenkins 1976). This model is most useful in situations where a time series of vital rate values is used to identify and fit a statistical model (*cf.* Lee 1974). Ecologists often prefer ARMA models in situations where serial autocorrelation over several time intervals is expected to be important.

The Semi-Markov Model: The possible values (states) of vital rates are as in the Markov case, but the time taken to make a transition from any one state to another in governed by a probability distribution which depends in general on both initial and final states. Here the history of the environment plays a stronger role.

The Catastrophe Model: This is a case of the I.I.D. model dealing with rare large environmental changes. One formulation is to suppose that in each time interval there is a very small probability of an event which will cause vital rates to reach extremely low levels; another formulation allows a probability distribution of times between successive catastrophic events, along with a distribution for the intensity of the catastrophic effect of vital rates. The biological view behind this model is of a population buffered against most small changes but vulnerable to large changes in environment.

Irrespective of the particular model used, we shall always assume that the random (environmental) process generating the vital rates converges towards an ergodic stationary state. *In general, we assume that the random process is in the stationary state*; for the approach to stationarity, see Tuljapurkar and Orzack (1980).

The next question is, what conditions apply to the possible values of the vital rates? The rates here are assumed to be always nonnegative, and in addition we assume **demographic weak ergodicity** (alternatively we assume that the values lie in an ergodic set). This guarantees that the dynamics of (4.1.1) are **stable** in the following sense. Note that we can rewrite (4.1.1) as an equation for the age structure,

$$\boldsymbol{Y}_{t+1} = \boldsymbol{X}_{t+1}\boldsymbol{Y}_t/(\boldsymbol{e}, \boldsymbol{X}_{t+1}\boldsymbol{Y}_t), \tag{4.1.2}$$

where we use the scalar product and $\boldsymbol{e}$ is a vector of 1s. The difference between (4.1.1) and (4.1.2) is that the $\boldsymbol{Y}$s, being vectors of proportions, are constrained so that $(\boldsymbol{e}, \boldsymbol{Y}) = 1$. Now in (4.1.2) pick two distinct initial structures $\boldsymbol{b}_0, \boldsymbol{c}_0$, and then apply the same random sequence of vital rates to both; call the resulting sequences of structure vectors $\boldsymbol{B}_t, \boldsymbol{C}_t$, respectively.

Then our ergodicity condition implies that $\boldsymbol{B}_t$ approaches $\boldsymbol{C}_t$ as t increases. This is stability, but of a special sort, since the age structures are stable towards a time-varying limit; *i.e.*, there is some sequence of structures $\widehat{\boldsymbol{Y}}_t$, say, and both $\boldsymbol{B}_t, \boldsymbol{C}_t$ approach $\widehat{\boldsymbol{Y}}_t$.

We now want to know: is there an analog to the stable age distribution? What is the asymptotic growth rate of population? What is the nature of convergence in the random model? Is there something like a reproductive value?

The answers to these questions are summarized in this chapter and further explored and applied in the rest of the book. First, however, we ask two basic questions. What features of (4.1.1) suggest that it will require a new theory? Why can we not apply insights derived from classical demography and random but non-structured models to get a handle on random demography? The answers to these questions lie in the facts that the dynamics of $\boldsymbol{N}$ are multiplicative and noncommutative, and in addition, the dynamics of $\boldsymbol{Y}$ are nonlinear. From (4.1.1) note that $\boldsymbol{N}$ is determined by a product of random (*i.e.*, randomly chosen) matrices, and that these matrices do not in general commute (*i.e.*, if we switch the order in which the matrices appear, the resulting product will change). It may be a good idea to convince oneself of this by, say, multiplying together two 2×2 Leslie matrices whose subdiagonal elements differ. From (4.1.2) note that the difference equation for $\boldsymbol{Y}$ is nonlinear and thus messier than (4.1.1) for $\boldsymbol{N}$. In classical demography this difference is irrelevant, but in the random case the moments of $\boldsymbol{Y}$ bear a complicated relationship to those of $\boldsymbol{N}$ and so the linear (4.1.1) does not shed much direct light on the nonlinear (4.2.2).

2 Results for Random Rates

There are many alternative models for random rates, so we cannot except a complete and universal theory. Instead we present results roughly in decreasing order of generality, alternating between stating mathematical results and interpreting them demographically.

2.1 General results

We begin with the most general set of assumptions:

Assumptions *4.2.1:*

(i) Demographic weak ergodicity holds in (2.2.1),

(ii) The random process generating vital rates is stationary and ergodic,

(iii) The logarithmic moment of vital rates is bounded,

$$\mathsf{E} \log_+ \|\boldsymbol{X}_1\| < \infty, \tag{4.2.1}$$

where E *indicates an expectation,* $\|\cdot\|$ *is any matrix norm, and* $\log_+(x) = \max\{0, \log x\}$.

Then we have (Furstenberg & Kesten 1960, Oseledec 1968, Cohen 1977a, Raghunathan 1979, Ruelle 1979):

(A) The long run growth rate of the logarithm of total population, or any part of population, is almost surely given by a number a independent of the initial population vector,

$$a = \lim_{t\to\infty} \left[\log(\boldsymbol{c}, \boldsymbol{N}_t)\right]/t \tag{4.2.2}$$

$$= \lim_{t\to\infty} (\log \|\boldsymbol{X}_t \boldsymbol{X}_{t-1} \dots \boldsymbol{X}_1\|)/t \tag{4.2.3}$$

$$= \lim_{t\to\infty} \{\mathsf{E} \log(\boldsymbol{c}, \boldsymbol{N}_t)\}/t, \tag{4.2.4}$$

where $\boldsymbol{c}$ is any vector of bounded nonnegative numbers.

(B) Starting from any initial structure $\boldsymbol{Y}_0$ the population converges to a (time dependent) stationary random sequence of structure vectors, $\widehat{\boldsymbol{Y}}_t$. This limiting sequence is independent of $\boldsymbol{Y}_0$.

(C) There is a stationary measure which describes the probability distribution of the joint sequence of vital rates and population structure vectors $\{\boldsymbol{X}_1, \widehat{\boldsymbol{Y}}_1, \boldsymbol{X}_2, \widehat{\boldsymbol{Y}}_2, \dots\}$.

(D) There are constants ρ_i for $i = 1$ through $i =$ (dimension of $\boldsymbol{N}$) such that

$$a = \rho_1 \geq \rho_2 \geq \dots . \tag{4.2.5}$$

The ρs are determined by the growth rates of exterior powers of the $\boldsymbol{X}$s, and are called Liapunov characteristic exponents . For example, let $\|\boldsymbol{x} \wedge \boldsymbol{y}\|$ be the volume of the parallelepiped spanned by vectors, $\boldsymbol{x}, \boldsymbol{y}$. Choose two nonproportional initial population vectors, say $\boldsymbol{b}_0, \boldsymbol{c}_0$, and apply (2.2.1) to produce two sequences of random vectors $\boldsymbol{B}_t$ and $\boldsymbol{C}_t$. Then the almost sure growth rate of the volume spanned by any two vectors is at most

$$a + \rho_2 = \max_{\{\boldsymbol{b}_0, \boldsymbol{c}_0\}} \lim_{t\to\infty} \log \|\boldsymbol{B}_t \wedge \boldsymbol{C}_t\|/t. \tag{4.2.6}$$

Similar results hold for sums of more exponents. (A notational point: Cohen (1977, etc.) writes $\log \lambda$ for the quantity a.)

We get an interesting general result if we add to assumptions (4.2.1) the

ASSUMPTION 4.2.2: *The random process generating vital rates can be run backwards in time, there being a unique time-reversed process which is stationary and ergodic.*

Then (Ruelle 1979) we have:

(E) Consider the adjoint (time-reversed) process associated with (4.1.2)

$$\boldsymbol{Z}_t = \boldsymbol{X}_t^T \boldsymbol{Z}_{t+1} / (\boldsymbol{e}, \boldsymbol{X}_t^T \boldsymbol{Z}_{t+1}), \tag{4.2.7}$$

where superscript T indicates a transpose. Suppose we fix a vector at time $t = t_1$, say $\boldsymbol{w} = \boldsymbol{z}_{t_1}$. Then (4.2.7) runs backwards through decreasing values of t, and we have that as $t \to -\infty$ the resulting vectors $\boldsymbol{Z}_t$ converge to a stationary random sequence of vectors $\widehat{\boldsymbol{Z}}_t$, say, independent of $\boldsymbol{w}_0$.

2.2 Interpretations

The central feature is that a is identified as the almost sure growth rate of population. It is also the average growth rate of the population. As the equations (4.2.2–4) show, the value of a is a function of the random properties of the rates. Note that a is also the average growth rate of any weighted sum of all or part of the population vector (from (4.2.4)).

Property (B) is the random rates counterpart of stability of population structure. Although there is a random limit to which the structures converge, there is no information of the properties of the limit. Property (C) expresses the strong overall convergence of rates plus structures to a statistical stationary state. Property (D) identifies an exponential convergence rate for population structures. All of these properties will take on substance when we become more explicit about the random process generating vital rates.

Property (E) identifies the stochastic analog of a reproductive value and helps to shed some light on the nature of reproductive value as a concept; more on this will follow.

2.3 Mixing and Lognormal Limit theorem

In addition to Assumptions (4.2.1), let us make:

Assumption *4.2.3: The random process generating vital rates is rapidly mixing.*

Technical aspects of the mixing condition are discussed by Furstenberg and Kesten (1960), Billingsley (1968), Tuljapurkar and Orzack (1980), and Heyde and Cohen (1985). Given such mixing, we have:

(F) Write $M_t = (\boldsymbol{e}, \boldsymbol{N}_t)$ for total population size at time t. There is some σ such that the asymptotic distribution of total population is lognormal,

$$\log\{(M_t - at)/\sigma\sqrt{t}\} \to N(0,1). \tag{4.2.8}$$

The quantity σ in (4.2.8) determines the asymptotic variance of the logarithm of population size. A limit theorem by Heyde and Cohen (1985) relevant to estimating σ will be discussed in a later chapter on forecasting and projection.

The significance of the lognormal distribution (4.2.8) originally was pointed out by Lewontin and Cohen (1969) for populations without age structure. Suppose that total population number follows the random growth equation

$$\begin{aligned} M_{t+1} &= R_{t+1} M_t \\ &= R_{t+1} R_t \ldots R_1 M_0. \end{aligned}$$

It is clear that $\log M_t$ will be asymptotically normally distributed with mean $t\,\mathsf{E}\,(\log R)$ and a variance which depends on the variance and correlation of the R_ts. *One* consequence of this is that the average growth rate, $(\log M_t/t)$ for large t, is in general going to be *less* than the growth rate of the average population, because

$$\log \mathsf{E}\,(M_t)/t \longrightarrow \log \mathsf{E}\,(R) \geq \mathsf{E}\,(\log R) = \mathrm{Limit}_{t\to\infty} \frac{1}{t} \mathsf{E}\, \log M_t.$$

This last inequality (Jensen's inequality, Karlin and Taylor 1975) is usually strict. A *second* consequence is that the distribution of M_t is left-skewed, indeed increasingly so for large t. Thus the most probable population sequences will tend asymptotically to fall below the average.

These conclusions all hold in the present case for structured populations. In particular the average growth rate a is the growth rate to be expected for a typical sample path, and it will in general be less than the growth rate of average population. The computation of a is considerably more involved than in the scalar case, but its qualitative importance is the same. See Tuljapurkar and Orzack (1980) for a fuller discussion and numerical examples.

2.4 Markovian rates

The key feature here is that statistical stationarity can be captured in a probability distribution function. Make

ASSUMPTION *4.2.4: The vital rates follow a countable state Markov process.*

Assumptions (4.2.1) and (4.2.3) are still in force. Then (Cohen 1977a,b, Lange 1979)

(G) There is a joint probability distribution of vital rates and population structures; call it $H(t, A, B) = \Pr\{\boldsymbol{X}_t \in A, \boldsymbol{Y}_t \in B\}$. Then as $t \to \infty$ this distribution converges to an equilibrium distribution, say $H^*(A, B)$, which reproduces itself under the action of (4.1.2).

(H) The average growth rate a can be computed as the average one-time-step growth rate. Note from (4.1.1–2) that

$$M_{t+1}/M_t = (\boldsymbol{e}, \boldsymbol{X}_{t+1}\boldsymbol{Y}_t), \tag{4.2.9}$$

so with M_t begin total population at time t, one has

$$a = \mathsf{E} \log(\boldsymbol{e}, \boldsymbol{X}_1 \boldsymbol{Y}_0) \tag{4.2.10}$$

with the average taken with respect to the stationary distribution H^*.

An equation for H^* can be written with some notational effort. Still more is known if we add

ASSUMPTION *4.2.5: The vital rates follow a finite state Markov process.*

Then (Bharucha 1961, Kushner 1966, Pollard 1969, Cohen 1977b, Tuljapurkar 1982a) one has:

(I) The moments of the population vector and its tensor powers, $\mathsf{E}\boldsymbol{N}_t$, $\mathsf{E}\boldsymbol{N}_t \otimes \boldsymbol{N}_t$, $\mathsf{E}\boldsymbol{N}_t \otimes \boldsymbol{N}_t \otimes \boldsymbol{N}_t \ldots$, can be computed explicitly as functions of time. Asymptotically these moments change geometrically with rates computable as dominant eigenvalues of explicitly known nonnegative matrices.

The final simplification is

ASSUMPTION *4.2.6: The random vital rates are determined by the* I.I.D. *model (see Section 4.1).*

Then we have:

(J) There is a probability distribution for the population structure vector, say $G(t, B) = \Pr\{\boldsymbol{Y} \in B\}$, and a corresponding stationary distribution $G^*(B)$ to which $G(t, B)$ converges as t increases.

Examples of G^* and H^* are given by Cohen (1977b), Tuljapurkar (1984), and in later chapters.

2.5 CONVERGENCE WITH RANDOM RATES

To understand the question of convergence, recall what happens if the population's projection matrix is fixed at $\boldsymbol{X}_t = \boldsymbol{b}$, as in Chapter 2. The age structure and growth rate then both approach their asymptotic values. The convergence questions (*cf.* Sections 2.2, 2.3) are: how rapidly does the growth rate approach r and $\boldsymbol{Y}_t$ approach $\boldsymbol{u}$? What oscillations in $\boldsymbol{N}_t$ are observed as $\boldsymbol{Y}_t \to \boldsymbol{u}$?

When vital rates are random the analogous questions concern random variables, and one has to ask how fast the average growth rate approaches

a, or how fast statistical distributions reach steady state. There are many aspects to convergence with random rates and it is not obvious how one ought to analyze convergence. I will describe below several natural measures of convergence along with their properties. Some of the results below are only applicable with Markov vital rates and are so labeled.

A. *Characteristic exponents*

Perhaps the most obvious approach to convergence is to seek a generalization of the subdominant eigenvalues of a fixed Leslie matrix. A partial generalization is found in the work of Oseledec (1968) on matrix products (see also Ruelle (1979) for a relevant discussion). Assuming stationary rates as described in Section 4.2.1, Oseledec's theorem says that there are k numbers called Liapunov characteristic exponents,

$$\rho_k \leq \rho_{k-1} \leq \cdots \leq \rho_1 = a, \tag{4.2.11}$$

of which only $m \leq k$ are distinct, and associated **random subspaces**

$$V_1 \subset V_2 \subset \cdots \subset V_m = R^k, \tag{4.2.12}$$

such that if we pick a random initial vector $\boldsymbol{C}_0 \in V_i/V_{i-1}$ and then compute $\boldsymbol{C}_t$ using (4.1.1) (with $V_0 = \{0\}$),

$$\rho_i = \lim_{t\to\infty} \frac{1}{t} \log(\boldsymbol{e}, \boldsymbol{C}_t), \tag{4.2.13}$$

where ρ_i is the appropriate exponent from (4.2.11). As (4.2.13) suggests, the quantities $\exp(\rho_i t)$ are asymptotically the singular values of the random product matrix $\boldsymbol{X}_t \boldsymbol{X}_{t-1} \cdots \boldsymbol{X}_1$.

To compute the ρs requires a different definition, based on Raghunathan (1979): choose j linearly independent vectors $\boldsymbol{w}_1(0), \ldots, \boldsymbol{w}_j(0)$ and apply (4.1.1) to each of these; then $a + \rho_2 + \cdots + \rho_j$ is maximum over all choices of the $\boldsymbol{w}$s of

$$\lim_{t\to\infty} \frac{1}{t} \log \|\boldsymbol{w}_1(t) \wedge \cdots \wedge \boldsymbol{w}_j(t)\| \tag{4.2.14}$$

where $\|\boldsymbol{w}_1 \wedge \cdots \wedge \boldsymbol{w}_j\|$ is the volume of the parallelepiped spanned by $\boldsymbol{w}_1, \ldots, \boldsymbol{w}_j$. For numerical use of this algorithm see Benettin et al. (1980). In the special case of (4.2.14) with $j = k$ one gets

$$\begin{aligned} a + \rho_2 + \cdots + \rho_k &= \lim_{t\to\infty} \frac{1}{t} \sum_{j=1}^{t} \log\left\{ \frac{\|\boldsymbol{X}_j \boldsymbol{w}_1(j-1) \wedge \cdots \wedge \boldsymbol{X}_j \boldsymbol{w}_k(j-1)\|}{\|\boldsymbol{w}_1(j-1) \wedge \cdots \wedge \boldsymbol{w}_k(j-1)\|} \right\} \\ &= \lim_{t\to\infty} \frac{1}{t} \sum_{j=1}^{t} \log|\det(\boldsymbol{X}_j)| = \mathsf{E} \log|\det(\boldsymbol{X}_1)|, \end{aligned}$$

where "det" means determinant. This result makes intuitive sense if one recalls that the ρs are related to singular values.

I will now show that ρ_2 is an analog to the quantity s = {log |first subdominant eigenvalue|} which describes convergence for a fixed Leslie matrix. Observe from (4.2.6) that if one picks any two vectors $\boldsymbol{w}_t(0)$, $\boldsymbol{w}_2(0)$, then $(a + \rho_2)$ gives the maximum possible asymptotic growth rate of the area spanned by

$$\boldsymbol{W}_1(t) = \boldsymbol{X}_t \boldsymbol{X}_{t-1} \cdots \boldsymbol{X}_1 \boldsymbol{w}_1(0), \tag{4.2.15}$$

and $\boldsymbol{W}_2(t)$ defined likewise. Now ergodicity means that as $t \to \infty$,

$$[\boldsymbol{W}_1(t)/\|\boldsymbol{W}_1(t)\| - \boldsymbol{W}_2(t)/\|\boldsymbol{W}_2(t)\|] \to 0.$$

Recalling the definitions of a and (4.2.14) it follows that

$$\begin{aligned}
&\lim_{t\to\infty} \frac{1}{t} \log \{\|(\boldsymbol{W}_1(t)/\|\boldsymbol{W}_1(t)\|) \wedge (\boldsymbol{W}_2(t)/\|\boldsymbol{W}_2(t)\|)\|\} \\
&= \lim_{t\to\infty} \left\{ \frac{1}{t} \log \|\boldsymbol{W}_t(t) \wedge \boldsymbol{W}_2(t)\| - \frac{1}{t} \log \|\boldsymbol{W}_1(t)| - \frac{1}{t} \log \|\boldsymbol{W}_t(t)\| \right\} \\
&\leq (\rho_2 + a) - 2a = (\rho_2 - a).
\end{aligned}$$

This result says that the initial difference between the **directions** of $\boldsymbol{W}_1(t)$, $\boldsymbol{W}_2(t)$ (*i.e.* the corresponding age-structures) goes to zero at the maximum rate $\exp[-(a - \rho_2)t]$ as $t \to \infty$. Clearly $(\rho_2 - a)$ is a convergence rate in the demographic sense. In general for nonproportional initial vectors $\boldsymbol{w}_1(0)$, $\boldsymbol{w}_2(0)$ chosen arbitrarily one expects that $\boldsymbol{W}_t(t) \wedge \boldsymbol{W}_2(t)$ will in fact change at the maximum rate. If one takes the matrices $\boldsymbol{X}$ to be fixed, the rate in (4.2.17) reduces to the classical convergence rate, as in Section 4.2.1.

Note however two important differences between the ρs and the higher eigenvalues of a fixed matrix. First the real ρs yield no information about the oscillatory transients. Second, the subspaces corresponding to the different ρs in (4.2.12) are random: they depend on the sequence of vital rates which will appear in (4.1.1), and it is not possible to specify them at a given initial time without knowledge of future rates. Thus the classical decay in components of age-structure orthogonal to a fixed stable structure does not generalize in any easy way. It is useful to compare the analysis of deterministically varying vital rates in Chapter 3 and by Kim and Sykes (1976) where similar questions arise.

The result (4.2.15) is particularly useful for Leslie matrices which have the form

$$\boldsymbol{X}_t = \begin{pmatrix} F_1(t) & \cdots & & F_k(t) \\ P_1(t) & \cdots & & 0 \\ \vdots & \ddots & & \vdots \\ 0 & \cdots & P_{k-1}(t) & 0 \end{pmatrix} \tag{4.2.16}$$

where Fs are fecundities and Ps are survival rates. For such matrices the result (4.2.15) simplifies to

$$a + \rho_2 + \cdots + \rho_k = \mathsf{E} \log(P_1 \cdots P_{k-1} F_k). \tag{4.2.17}$$

In the $k = 2$ case this yields ρ_2 directly.

B. *Other approaches to measuring convergence rate*

One approach to measuring the convergence rate is the construction of a measure on the sequence space $\{(\boldsymbol{X}_1, \boldsymbol{Y}_1), (\boldsymbol{X}_2, \boldsymbol{Y}_2), \ldots\}$. The random process of vital rates can be described by a measure on the projected sequences $\{\boldsymbol{X}_1, \boldsymbol{X}_2, \ldots\}$ and one can view the bivariate sequence space as a skew-product dynamical system (Abramov and Rohlin 1967). Ruelle (1978, Corollary 6.23) and Furstenberg and Kesten (1960) consider the construction of a suitable measure on the bivariate sequence space. One can now generalize the results in Tuljapurkar (1982c) to show that the entropy of this skew-product system provides a lower bound on the rate at which the equilibrium measure is approached in sequence space.

Another approach is to examine the distributional convergence of total numbers scaled to $\log|\boldsymbol{N}_t|/\sqrt{t}$ as in (4.2.8). A numerically based analysis for Markov vital rates is given in Tuljapurkar and Orzack (1980).

The situation is much clearer with Markov rates (Cohen 1977a, b). Here the state of the pair $\{\boldsymbol{X}_t, \boldsymbol{Y}_t\}$ can be represented as a point in a (product) space of $k^2 + (k-1)$ dimensions (matrix plus normalized vector). Suppose that μ is a measure on this space which is reproduced under the action of (4.1.1). Then it obeys an equation of the form

$$\mu(dz) = \int K(dz, v)\mu(dv), \tag{4.2.18}$$

where K is a nonnegative transition kernel. In general one expects a family of eigensolutions of the equation

$$\lambda\, \phi_\lambda(dz) = \int K(dz, v)\phi_\lambda(dv). \tag{4.2.19}$$

From the theory of positive operators (Schaefer 1974) one has $|\lambda| \leq 1$; $\lambda = 1$ corresponds to the stationary distribution of Property G in Section 4.2.4. Thus a natural convergence rate is given by the eigenvalue λ for which $|\lambda|$ closest to 1.

A final approach is to ask how rapidly moments of $\boldsymbol{N}_t$ approach their asymptotic growth trends. With Markov rates it is known (Tuljapurkar 1982a; see also Chapter 7) that

$$\mathsf{E}(\boldsymbol{N}_t \otimes \boldsymbol{N}_t \otimes \cdots \otimes \boldsymbol{N}_t) \sim (\text{Constants}) \cdot \beta_\ell^t + (\text{Constants}) \cdot \gamma_\ell^t, \tag{4.2.20}$$

with β_ℓ and γ_ℓ, $|\gamma_\ell| < \beta_\ell$, explicitly computable as eigenvalues of a particular matrix. Therefore one has a family of convergence rates given by $|\gamma_\ell|/\beta_\ell$. In the above paper β_ℓ, γ_ℓ are studied for the case when $\|\boldsymbol{X}_t - \mathsf{E}(\boldsymbol{X}_t)\|$ is small and the results suggest that the ratios $|\gamma_\ell|/\beta_\ell$ decline rapidly with increasing ℓ. Thus $|\gamma_1|/\beta_1$ may be an adequate index of convergence for the moments. Note that when the matrices $\boldsymbol{X}_t$ are independently and identically distributed this ratio reduces to the classical convergence rate.

2.6 Simulations

Useful insight into the consequences of random rates has come from numerical simulations, *e.g.*, Boyce (1977), Cohen (1977b), Tuljapurkar and Orzack (1980), Slade and Levenson (1982). Pollard (1973) describes a way of simulating the more complex case of populations with "demographic" stochasticity added in. There is a large and relevant literature on simulation methodology; the book by Ripley (1987) is a concise introduction. It is fair to say that simulations are most effective when informed by theoretical reasoning, and when their potential limitations are kept carefully in view.

3 Assessing the Results

We have presented above the skeleton of a fairly general theory for random rates. However, these results bear a distant relationship to the substantive problems listed in the introduction, in the same way that the Perron-Frobenius theorem is not classical demography. In addition to the obstacles posed by the evident complexity of random rates theory, some theoretical issues remain unresolved. The nature of oscillatory transients in the random theory remains unclear, although Lee (1974) has discussed some of their properties and their significance in population dynamics. The significance of the reproductive value has not been explored. There is limited information about the functional dependence of objects like a and the ρs on the properties of the underlying vital rates.

This book addresses these complexities by studying stylized and practical examples, approximate and exact analytical results, and evolutionary models. The goal is to resolve aspects of the relation between the general theory and applications, develop insight into the consequences of random vital rates, and partially fill some gaps in the theory. The next chapter summarizes examples of populations for which this theory is ideally suited.

5

EXAMPLES

There is increasing interest in the impact of environmental variability on vital rates. I summarize below examples in which such variation has actually been measured. The examples include plants and animals and point to a growing recognition that such data are needed and useful. However, most of the data are limited in their sampling both of environments and of the transition structure between environments.

1 Human Population Projection

Demographers of humans have always been in the vanguard of studies of variation. There are some fairly long time series of vital rates, mostly for Western countries in the 20th century. All of these are typically available in the standard Leslie matrix form, often in dimension 10×10 (or greater) with age classes 5 years wide. A particularly notable series for Sweden was utilized by both Kim and Sykes (1976) and Cohen (1986). There has also been considerable work on the use of time series models to describe either vital rates (Lee 1974, 1977) or the birth sequence itself (Saboia 1977, MacDonald 1979). I use Lee's approach in later chapters of this book, treating vital rates as determined by an autoregressive moving-average (ARMA) model (Box and Jenkins 1970).

2 Large Mammals: Elk and Ungulates

Wallace (1986) reports on a population of Rocky Mountain Elk (*Cervis elephus nelsoni*) in Washington state. He used several years of population counts and data on climatic variables which affected vital rates. His final population model uses 3 year classes and a 3×3 Leslie projection matrix in which only the survival rate of the age class is stochastic. This stochasticity is modeled by a discrete I.I.D. random sequence with a large number of states.

A related study by Van Sickle (1989) considers vital rates broadly applicable to African ungulates. In agreement with Wallace on elk, he argues that most variability occurs in survival rate of young, and he makes rough estimates of the extent of variation.

3 Long-Lived Fish

Fisheries analysts have long had to grapple with substantial temporal variation in estimates of recruitment into fisheries. A classical study of the consequences of such variation was done by Murphy (1968). His life histories were based on extensive work with the Pacific sardine , but the structure of his stochastic projection model applies to many species including the American shad (Leggett and Carscadden 1978) and the striped bass (Cohen *et al.* 1983). The model is an age classified Leslie projection matrix in which only the survival of the youngest age class is stochastically variable. It is interesting to note the similarity to models for large mammals (Sec. 5.2); however, the age patterns of average age-specific reproduction and survival are quite different. A point worth mentioning for the egg-laying fish of the sort Murphy considers is that their numerical average fecundity is often enormous (10^4–10^6) while the numerical average survival rate of very young fish is tiny (10^{-4}–10^{-6}). However, this pair of scale factors is easily factored out of the projection scheme, as Murphy implicitly does in his paper.

4 Plants: Biennials

Klinkhamer and de Jong (1983), Roerdink (1987), and de Jong and Klinkhamer (1988) discuss the evolution of delayed flowering in monocarpic perennials. Examples of the species they consider are the spear thistle *(Cirsium vulgare)* and hound's-tongue *(Cynoglossum officianale)*. Both species are biennials in the sense that plants flower in the second year or later and die after flowering. However, plants of both species can delay flowering, often by several years. Klinkhamer and de Jong (1983) argued that delayed flowering was an adaptation to variability in reproductive success, which includes seed production, seed survival and germination, and seedling survival. Some estimates of this variability are given in de Jong and Klinkhamer (1988). Roerdink (1987) uses the theory discussed in this book to analyze a formal model of delayed flowering; I discuss his results in a later chapter. The projection matrices here are age-structured.

5 Plants: Perennials

A more complex projection analysis is given by Bierzychudek (1982) for the demography of Jack-in-the-pulpit, *(Arisaema triphyllum)*, a long-lived forest herb. Her studies suggested that *A. triphyllum* is best described by a size-classified model (there are no usable indicators of age in the field). The seven size classes used were: 1, seeds; 2 to 7, leaf area classes. The resulting 7×7 projection matrices were structured as follows. Elements in the first row (b_{1i}, $i = 1, \ldots, 7$) describe seed production by plants in successive size classes. Leading subdiagonal elements (b_{21}, b_{32}, etc.) are transition rates from one size class to the next largest. Superdiagonal elements in rows 2-7 (such as b_{23}, b_{24}, etc.) are transition rates from large to smaller sizes. Diagonal elements are transition rates which give the probability of staying in the same size class. There was variation both within spatial locations over time and between spatial locations. Of the transitions that are not observed in an age-structured population, the particularly novel ones here are the transitions into size class 2 from larger classes. These transitions include a component due to **clonal reproduction**, in which larger plants produce new individuals vegetatively. Thus the second row ($b_{2j}, j > 2$) of the projection matrix contains the asexual reproductive rates for the population.

Yet another example of a perennial plant is provided by Moloney's (1988) study of the perennial grass *Danthonia sericea*. Five spatial locations were tracked over 2 years to obtain a total of 10 size-classified projection matrices. There were 6 size categories according to the number of leaves per individual (1–2, 3–6, 7–13, 14–27, 28–56, >56). The structure of these matrices is exemplified by the following matrix (the June 1983–June 1984, location A matrix from Moloney (1988), Table 2). The first row includes all contributions from seeds to newly germinated individuals *plus* transitions from larger size classes into size class 1. All other rows describe transitions between existing individuals. The matrix is

$$\begin{pmatrix} 0.21 & 0.47 & 1.89 & 6.57 & 15.10 & 21.75 \\ 0.19 & 0.55 & 0.10 & 0 & 0 & 0 \\ 0.02 & 0.19 & 0.57 & 0 & 0 & 0 \\ 0 & 0.02 & 0.27 & 0.77 & 0.30 & 0 \\ 0 & 0 & 0 & 0.23 & 0.60 & 0 \\ 0 & 0 & 0 & 0 & 0.10 & 1.00 \end{pmatrix}.$$

The first row, especially the last four entries, are dominated by seed production. These entries show tremendous variability from year to year, and also between sites.

6 Plants: Morphological Structure

A strikingly different example comes from the work of Huenneke and Marks (1987) and Huenneke (1986) on a clonal shrub, the speckled alder *(Alnus incana)*. This alder forms dense thickets where seedling reproduction is virtually absent and growth primarily vegetative. Huenneke argues that the dynamics of a local population (of one or a few thickets) can therefore be represented by the size and number of stems in the thickets. A size structured model was constructed in which the classes are distinguished by stem diameter at breast height. A series of censuses was used to construct transition rates between classes; considerable spatial and temporal variability was observed in these rates.

7 Plants: Age and Size

My final example is from Van Groenendael and Slim (1988) who develop projection matrices for the iterocarpic perennial *Plantago lanceolata.* They argue that neither a purely age structured nor a purely size structured (Werner and Caswell 1977) model is appropriate. Instead they use a joint classification into age and size categories following Goodman (1969) and Law (1983). The result is a complex projection matrix with as many as 90 discrete classes. The same paper discusses temporal and environmental variation in the transition rates between classes and attempts to draw some conclusions about the effect of random variation on population extinction.

6

ESS AND ALLELE INVASION

Population biologists have long been concerned with the evolution of demographic vital rates, and so with the analysis of combined genetic-demographic models. One approach to such models is to study ESS (evolutionary stable state) criteria (e.g., Charlesworth 1980): consider a genetic model in which two alleles at a single diploid locus affect the life history phenotypes in a randomly mating population. Suppose the population is initially homozygous for one allele and introduce the other allele at low frequency. Then ask: what conditions determine the invasion (increase in frequency) of the rare allele? The answer identifies the natural "fitness" measure for a discussion of the evolution of life histories. The analysis for sexually reproducing populations is different from that in which clonal reproduction is also possible, and I treat these separately below.

1 Sexually Reproducing Populations

I restrict myself first to populations in which there is no clonal (vegetative) reproduction. The population genetic model I use is based on Charlesworth (1980) and assumes a fixed sex-ratio, no sex differences, and random mating. The population is divided into discrete classes and class 1 consists of newly born individuals. The projection matrices for the population have the form

$$\boldsymbol{X}_t = \begin{pmatrix} \boldsymbol{F}_t^T \\ \boldsymbol{S}_t \end{pmatrix} \tag{6.1.1}$$

where the vector $\boldsymbol{F}_t$ contains birth rates and the matrix $\boldsymbol{S}_t$ contains all other transitions between population classes. Consequently, products of the form

$$\begin{pmatrix} \boldsymbol{0}^T \\ \boldsymbol{S}_t \end{pmatrix}\begin{pmatrix} \boldsymbol{0}^T \\ \boldsymbol{S}_{t-1} \end{pmatrix}\cdots\begin{pmatrix} \boldsymbol{0}^T \\ \boldsymbol{S}_1 \end{pmatrix} \longrightarrow (\boldsymbol{0})$$

as t increases.

A life history is determined by the genotype at a diploid locus. Assuming alleles A_1, A_2, the genotype A_iA_j $(i,j = 1,2)$ is described by a population vector $\boldsymbol{N}_{ij,t} = \boldsymbol{N}_{ji,t}$ and determines a life history phenotype consisting of a random sequence of growth matrices $\boldsymbol{X}_{ij,t}$. The total number of newborn individuals at time t is

$$B_t = \sum_{i,j} \boldsymbol{N}_{ij,t}(1).$$

We need only track gene frequencies $P_{i,t}$ among newborns.

The population and genetic dynamics follow the recursions:

$$\begin{aligned} B_{t+1} &= \sum_{i,j} (\boldsymbol{X}_{ij,t+1}\boldsymbol{N}_{ij,t})(1), \\ N_{i,j,t+1} &= P_{i,t+1}P_{j,t+1}B_{t+1}, \\ 2B_{t+1}P_{i,t+1} &= \sum_{j} (\boldsymbol{X}_{ij,t+1}\boldsymbol{N}_{ij,t} + \boldsymbol{X}_{ji,t+1}\boldsymbol{N}_{ji,t})(1), \\ N_{ij,t+1}(m+1) &= (\boldsymbol{X}_{ij,t+1}\boldsymbol{N}_{ij,t})(m) \end{aligned} \tag{6.1.2}$$

with $i, j = 1, 2$ and $m = 1, \ldots, k-1$. As usual the bracketed integers on the far right indicate vector components.

Initially assume only allele A_1 is present so population growth follows

$$\boldsymbol{Q}_{11,t+1} = \boldsymbol{X}_{11,t+1}\boldsymbol{Q}_{11,t} \tag{6.1.3}$$

with $\boldsymbol{N}_{11} = \boldsymbol{Q}_{11}$ and all other $\boldsymbol{N}_{ij} = \mathbf{0}$. Introducing allele A_2 at low frequency means that $\boldsymbol{N}_{11,t} = \boldsymbol{Q}_{11,t} + \varepsilon_{11,t}$ and $\boldsymbol{N}_{ij,t} = \varepsilon_{ij,t}$ where the $\|\varepsilon_t\|$ are small. Linearize the system (6.1.2) around the time dependent steady state (6.1.3) to get

$$\begin{pmatrix} \varepsilon_{11} \\ \varepsilon_{12} \\ \varepsilon_{21} \\ \varepsilon_{22} \end{pmatrix}_{t+1} = \begin{pmatrix} \boldsymbol{X}_{11} & \mathbf{0} & \mathbf{0} & \begin{pmatrix} -\boldsymbol{F}_{22} \\ \mathbf{0} \end{pmatrix} \\ \mathbf{0} & \frac{1}{2}\boldsymbol{X}_{12} & \frac{1}{2}\boldsymbol{X}_{21} & \begin{pmatrix} \boldsymbol{F}_{22} \\ \mathbf{0} \end{pmatrix} \\ \mathbf{0} & \frac{1}{2}\boldsymbol{X}_{12} & \frac{1}{2}\boldsymbol{X}_{21} & \begin{pmatrix} \boldsymbol{F}_{22} \\ \mathbf{0} \end{pmatrix} \\ \mathbf{0} & \mathbf{0} & \mathbf{0} & \begin{pmatrix} \mathbf{0} \\ \boldsymbol{S}_{22} \end{pmatrix} \end{pmatrix}_{t+1} \begin{pmatrix} \varepsilon_{11} \\ \varepsilon_{12} \\ \varepsilon_{21} \\ \varepsilon_{22} \end{pmatrix}_t . \tag{6.1.4}$$

Asymptotically (6.1.4) shows that $\varepsilon_{22,t} \to \mathbf{0}$ (recall the discussion following (6.1.1)) and so we need only consider heterozygotes. Write $\varepsilon = \varepsilon_{12} + \varepsilon_{21}$ and see from (6.1.4) that for large t

$$\varepsilon_{t+1} = \boldsymbol{X}_{12,t+1}\varepsilon_t. \tag{6.1.5}$$

The frequency of allele A_2 can be written to first order in ε by using the third equation of (6.1.2) to get

$$P_{2,t+1} = \frac{\varepsilon_{t+1}(1)}{2Q_{11,t+1}(1)} + \frac{\varepsilon_{22,t+1}(1)}{Q_{11,t+1}(1)}. \tag{6.1.6}$$

Since $\varepsilon_{22,t} \to 0$ at large t, the rate of change of $P_{2,t}$ is given by

$$\begin{aligned} \frac{1}{t}\log P_{2,t+1} &= \frac{1}{t}\log \varepsilon_{t+1}(1) - \frac{1}{t}\log Q_{11,t+1}(1) \\ &\to (a_{12} - a_{11}) \quad \text{as} \quad t \to \infty. \end{aligned} \tag{6.1.7}$$

Here a_{12} is the growth rate of a hypothetical population composed of individuals whose projection matrices are $\boldsymbol{X}_{12,t}$, and a_{11} is the growth rate of a population homozygous for allele A_1.

Thus we conclude that a is the "natural" fitness measure in a random environment. This result is the (multidimensional) demographic version of the classical results for genes under viability selection (Haldane and Jayakar 1963, Karlin and Lieberman 1975). It is obvious from (6.1.7) that a "protected" polymorphism of A_1 and A_2 is obtained if

$$a_{11} < a_{21} > a_{22}, \tag{6.1.8}$$

the familiar condition of heterozygote superiority. It is clear that for life histories of the structure considered here (i.e., no cloning), the growth rate a *plays the same role for random rates as does the classical* r *in deterministic evolutionary theory.*

2 Clonal Reproduction

I now consider a more complicated situation, exemplified by Bierzychudek's description of jack-in-the-pulpit (Section 5.5). The key difference is that there is clonal reproduction, so there are now (at least) two population classes which contain newly produced individuals. Let class 1 remain the class of sexually produced newborn individuals, and let class 2 (as in Bierzychudek's example) include all newly produced clones. Then a decomposition of the projection matrix analogous to (6.1.1) can be written

$$\boldsymbol{X}_t = \begin{pmatrix} \boldsymbol{F}_t^T \\ \boldsymbol{S}_t \end{pmatrix} = \begin{pmatrix} \boldsymbol{F}_t^T \\ \boldsymbol{C}_t^T \\ \boldsymbol{R}_t \end{pmatrix} \tag{6.2.1}$$

Here $\boldsymbol{F}_t$ contains sexual reproduction rates, $\boldsymbol{C}_t$ contains clonal reproduction rates, and $\boldsymbol{R}_t$ contains all other transitions. So it is no longer true that products of the type

$$\begin{pmatrix} \boldsymbol{0}^T \\ \boldsymbol{S}_t \end{pmatrix} \begin{pmatrix} \boldsymbol{0}^T \\ \boldsymbol{S}_{t-1} \end{pmatrix} \cdots \begin{pmatrix} \boldsymbol{0}^T \\ \boldsymbol{S}_1 \end{pmatrix}$$

will go to zero as t increases.

To proceed with an allelic invasion analysis, we can still use equations (6.1.2) with the understanding that B_t counts only sexually produced offspring. Linearizing as in (6.1.4) we find the pair of equations

$$\epsilon_{12,t+1} = \boldsymbol{X}_{12,t+1}\epsilon_{12,t} + \begin{pmatrix} \boldsymbol{F}_{22,t+1} \\ \boldsymbol{0} \end{pmatrix} \epsilon_{22,t}, \tag{6.2.2}$$

$$\epsilon_{22,t+1} = \begin{pmatrix} \boldsymbol{0}^T \\ \boldsymbol{S}_{22,t+1} \end{pmatrix} \epsilon_{22,t}.$$

In contrast to the situation described following (6.1.4), note that $\varepsilon_{22,t}$ *does not* necessarily go to zero over time, because the $\boldsymbol{S}_{22}$s contain cloning rates which can be substantial. The invasion dynamics of allele A_2 depend on the cloning rate and the difference between a_{12} and a_{11} is *not* a sufficient criterion to determine invasion.

7

MOMENTS OF THE POPULATION VECTOR

The basic population equation (4.1.1) is linear and this linearity ought to be of some use. This chapter shows that for both the I.I.D. and the (finite) Markov models, the linearity allows us to compute moments of the population vector.

1 Serially Independent Environments

In the basic model

$$\boldsymbol{N}_{t+1} = \boldsymbol{X}_{t+1}\boldsymbol{N}_t, \tag{7.1.1}$$

suppose that $\boldsymbol{N}_0 = \boldsymbol{n}_0$ is a fixed vector and that successive $\boldsymbol{X}_t$s are independently randomly generated with the *same* probability distribution. Then the average population vector is given by

$$\begin{aligned} \mathsf{E}(\boldsymbol{N}_t) &= \mathsf{E}(\boldsymbol{X}_t\boldsymbol{N}_{t-1}) \\ &= \mathsf{E}(\boldsymbol{X}_t)\mathsf{E}(\boldsymbol{N}_{t-1}). \end{aligned} \tag{7.1.2}$$

The second line of (7.1.2) follows because $\boldsymbol{X}_{t+1}$ is independent of the previous history of the population. Letting

$$\mathsf{E}(\boldsymbol{X}_t) = \boldsymbol{b}, \tag{7.1.3}$$

we see that the vector $\mathsf{E}(\boldsymbol{N}_t)$ grows asymptotically at the rate μ^t where μ is the dominant eigenvalue of $\boldsymbol{b}$. Thus $\log \mu$ is the long run growth rate of average population. It *does not* follow from (7.1.2) that the average age structure, $\mathsf{E}(\boldsymbol{N}_t/M_t)$, has *any* relationship to the right eigenvector of $\boldsymbol{b}$ corresponding to μ. This latter fact should be obvious from the nonlinearity of (4.1.2) compared with (4.1.1).

The computation of higher moments (Pollard 1966) rests on a remarkable property of the Kronecker product (2.3.5), namely, that for matrices $\boldsymbol{b}, \boldsymbol{c}$ and vectors $\boldsymbol{x}, \boldsymbol{y}$,

$$\boldsymbol{b}\boldsymbol{x} \otimes \boldsymbol{c}\boldsymbol{y} = (\boldsymbol{b} \otimes \boldsymbol{c})(\boldsymbol{x} \otimes \boldsymbol{y}). \tag{7.1.4}$$

From (7.1.1) and (7.1.4) it follows that

$$\boldsymbol{N}_{t+1} \otimes \boldsymbol{N}_{t+1} = (\boldsymbol{X}_{t+1} \otimes \boldsymbol{X}_{t+1})(\boldsymbol{N}_t \otimes \boldsymbol{N}_t), \tag{7.1.5}$$

$$\boldsymbol{N}_{t+1} \otimes \boldsymbol{N}_{t+1} \otimes \boldsymbol{N}_{t+1} = (\boldsymbol{X}_{t+1} \otimes \boldsymbol{X}_{t+1} \otimes \boldsymbol{X}_{t+1})(\boldsymbol{N}_t \otimes \boldsymbol{N}_t \otimes \boldsymbol{N}_t), \tag{7.1.6}$$

and so forth. The components of $\boldsymbol{N}_t \otimes \boldsymbol{N}_t$ are the products $N_t(i)N_t(j)$, $i, j = 1, \ldots, k$, so $\mathsf{E}(\boldsymbol{N}_t \otimes \boldsymbol{N}_t)$ describes all second moments of the population vector. In addition, the second moment of total population

$$\mathsf{E}(M_t^2) = \mathsf{E}(\boldsymbol{e}, \boldsymbol{N}_t)^2 \tag{7.1.7}$$

$$= \mathsf{E}(\boldsymbol{e} \otimes \boldsymbol{e},\ \boldsymbol{N}_t \otimes \boldsymbol{N}_t). \tag{7.1.8}$$

From (7.1.5) the second moments follow the recursion

$$\mathsf{E}(\boldsymbol{N}_{t+1} \otimes \boldsymbol{N}_{t+1}) = \mathsf{E}(\boldsymbol{X}_{t+1} \otimes \boldsymbol{X}_{t+1})\mathsf{E}(\boldsymbol{N}_t \otimes \boldsymbol{N}_t), \tag{7.1.9}$$

and the long run growth rate of these second moments is $\log \beta$, where β is the dominant eigenvalue of $\mathsf{E}(\boldsymbol{X}_t \otimes \boldsymbol{X}_t)$. We can go on to higher moments in (7.1.6).

2 Markovian Environments

In (7.1.1) assume that only $r(\geq 2)$ different matrices can appear. These matrices $\{\boldsymbol{b}_i, i = 1, \ldots, r\}$ form a set and transitions between the matrices follow a Markov chain. Thus there are probabilities

$$p_{ij} = \Pr\{\boldsymbol{X}_{t+1} = \boldsymbol{b}_j \mid \boldsymbol{X}_t = \boldsymbol{b}_i\}, \tag{7.2.1}$$

and they can be written as a transition probability matrix $\boldsymbol{p} = (p_{ij})$. There is a remarkable formula due to Bharucha (1961), reported in the literature on stochastic models by Kushner (1966), which allows the population moments to be computed for this Markovian case. Cohen (1977b) rediscovered this formula for the first moment; Tuljapurkar (1982a) applied Bharucha's general formula to the first and second moments.

The derivation of the formula is instructive and I will sketch it here, rather than pulling the result out of thin air. Suppose that at time $t = 1$, the environmental distribution is given by

$$q_i = \Pr\{\boldsymbol{X}_0 = \boldsymbol{b}_i\}, \quad i = 1, \ldots, r, \tag{7.2.2}$$

and that the population vector is some fixed $\boldsymbol{n}_0 = \boldsymbol{N}_0$. From time $t = 1$ to time $t > 1$, let the sequence of environments which appears in (7.1.1) according to the transition probabilities (7.2.2) be $w = (i_t, i_{t-1}, \ldots, i_1)$. Then

$$\boldsymbol{N}_t(w) = \boldsymbol{b}_{i_t}\boldsymbol{b}_{i_{t-1}} \ldots \boldsymbol{b}_{i_1}\boldsymbol{n}_0, \tag{7.2.3}$$

and

$$\mathsf{E}\boldsymbol{N}_t = \sum_w \Pr\{w\}\boldsymbol{N}_t(w)$$

$$= \sum_w p_{i_{t-1}i_t} \ldots p_{i_1 i_2} q_{i_1} \boldsymbol{b}_{i_t}\boldsymbol{b}_{i_{t-1}} \ldots \boldsymbol{b}_{i_1}\boldsymbol{n}_0. \tag{7.2.4}$$

Now introduce the transpose matrix

$$\boldsymbol{p}^T = (p'_{ij}) = (p_{ji}), \tag{7.2.5}$$

and note that (7.2.4) can be written as

$$\sum_{(i_1 i_2 \ldots i_t)} \boldsymbol{d}_{i_t i_{t-1}} \boldsymbol{d}_{i_{t-1} i_{t-2}} \ldots \boldsymbol{d}_{i_2 i_1} b_{i_1} q_{i_1} \boldsymbol{n}_0 \tag{7.2.6}$$

where

$$\boldsymbol{d}_{ij} = \boldsymbol{b}_i p'_{ij}. \tag{7.2.7}$$

If we imagine a matrix whose elements are $\boldsymbol{d}_{ij}$, the sum in (7.2.6) resembles closely an element of a power of this matrix. Write

$$\boldsymbol{d} = (\boldsymbol{d}_{ij}) = \begin{pmatrix} \boldsymbol{d}_{11} & \cdots & \boldsymbol{d}_{1r} \\ \vdots & & \\ \boldsymbol{d}_{r1} & \cdots & \boldsymbol{d}_{rr} \end{pmatrix}, \tag{7.2.8}$$

a matrix of $k \times k$ blocks. Then $\boldsymbol{d}^{t-1}$ is also a matrix of $k \times k$ blocks and its α, β block element is

$$(\boldsymbol{d}^{t-1})_{\alpha\beta} = \sum_{(i_2 \ldots i_{t-1})} \boldsymbol{d}_{\alpha i_{t-1}} \boldsymbol{d}_{i_{t-1} i_{t-2}} \ldots \boldsymbol{d}_{i_2 \beta}. \tag{7.2.9}$$

Using this we have

$$\mathsf{E}\,\boldsymbol{N}_t = \sum_{\alpha,\beta} (\boldsymbol{d}^{t-1})_{\alpha\beta}\, q_\beta \boldsymbol{b}_\beta \boldsymbol{n}_0. \tag{7.2.10}$$

The sums in (7.2.10) are achieved by defining

$$\begin{aligned}
\boldsymbol{I} &= \begin{pmatrix} 1 & & \\ & 1 & 0 \\ & & \ddots & \\ 0 & & 1 \end{pmatrix} \\
&= \mathrm{diag}(1, 1, \ldots, 1) = \quad \text{the } k \times k \text{ identity matrix}, && (7.2.11) \\
\boldsymbol{f} &= \mathrm{diag}(\boldsymbol{b}_1, \boldsymbol{b}_2, \ldots, \boldsymbol{b}_r), && (7.2.12) \\
\boldsymbol{J} &= \mathrm{diag}\,\underset{\leftarrow r \text{ blocks} \rightarrow}{(\,\boldsymbol{I}, \boldsymbol{I} \ldots, \boldsymbol{I}\,)}, && (7.2.13) \\
\boldsymbol{z}^T &= (q_1 \boldsymbol{I}, q_2 \boldsymbol{I}, \ldots, q_r \boldsymbol{I}). && (7.2.14)
\end{aligned}$$

Then

$$\mathsf{E}\,\boldsymbol{N}_t = \boldsymbol{J}(\boldsymbol{d})^{t-1} \boldsymbol{f} \boldsymbol{z} \boldsymbol{n}_0. \tag{7.2.15}$$

A final convenient notational step is to note from (7.2.7), (7.2.8), and (7.2.11)–(7.2.12) that

$$\boldsymbol{d} = \boldsymbol{f}(\boldsymbol{p}^T \otimes \boldsymbol{I}). \tag{7.2.16}$$

The striking aspect of the above method is that it works equally well for the equations giving $\boldsymbol{N}_t \otimes \boldsymbol{N}_t$, $\boldsymbol{N}_t \otimes \boldsymbol{N}_t \otimes \boldsymbol{N}_t$, etc. The reader can show, for example, that

$$\mathsf{E}\,\boldsymbol{N}_t \otimes \boldsymbol{N}_t = \boldsymbol{J}_2(\boldsymbol{d}_2)^{t-1}\boldsymbol{f}_2\boldsymbol{z}_2\boldsymbol{n}_0 \otimes \boldsymbol{n}_0, \tag{7.2.17}$$

where we use the matrices

$$\boldsymbol{J}_2 = \underset{\longleftarrow\ r\text{blocks, each } k^2 \times k^2\ \longrightarrow}{\begin{pmatrix} \boldsymbol{I}\otimes\boldsymbol{I} & \boldsymbol{I}\otimes\boldsymbol{I} & \cdots & \boldsymbol{I}\otimes\boldsymbol{I} \end{pmatrix}} \tag{7.2.18}$$

$$\boldsymbol{f}_2 = \begin{pmatrix} \boldsymbol{b}_1\otimes\boldsymbol{b}_1 & & & \\ & \boldsymbol{b}_2\otimes\boldsymbol{b}_2 & & \mathbf{0} \\ & & \ddots & \\ \mathbf{0} & & & \boldsymbol{b}_r\otimes\boldsymbol{b}_r \end{pmatrix}, \tag{7.2.19}$$

$$\boldsymbol{z}_2 = \begin{pmatrix} q_1\boldsymbol{I}\otimes\boldsymbol{I} \\ \vdots \\ q_r\boldsymbol{I}\otimes\boldsymbol{I} \end{pmatrix}, \tag{7.2.20}$$

$$\boldsymbol{d}_2 = \boldsymbol{f}_2(\boldsymbol{p}^T \otimes \boldsymbol{I} \otimes \boldsymbol{I}). \tag{7.2.21}$$

To summarize, the moments of the population vector can be computed explicitly using the Bharucha formulae such as (7.2.15) and (7.2.17). Furthermore, the matrices $\boldsymbol{d}$ and $\boldsymbol{d}_2$ in (7.2.16) and (7.2.21) are nonnegative and primitive (by our assumptions of demographic ergodicity). Therefore, the long run growth rates of $\mathsf{E}\,\boldsymbol{N}_t$ and $\mathsf{E}\,\boldsymbol{N}_t \otimes \boldsymbol{N}_t$ are $\log\mu$ and $\log\beta$ respectively, where μ and β are the dominant eigenvalues of $\boldsymbol{d}$ and $\boldsymbol{d}_2$.

The problems with the results given in this section are:

- it would be much more useful to compute a and σ^2, the growth rate and variance of $\log M_t$ (see 4.2.8);
- the matrices $\boldsymbol{d}$ and $\boldsymbol{d}_2$ have complicated structure and the determinants of μ and β are hard to discern.

The next three sections consider ways in which one can partially address these problems.

3 Inequalities and Exact Results

Cohen (1979 a,b) and Tuljapurkar (1982a) present some inequalities and related results concerning μ, β and a. I begin with the obvious fact that since variances are nonnegative, we must have

$$\beta \geq \mu^2. \tag{7.3.1}$$

We know already (Section 4.2.3) that $a \leq \log \mu$ so

$$a \leq \log \mu \leq \frac{1}{2} \log \beta. \tag{7.3.2}$$

Recall now (Seneta 1981) that the smallest (resp. largest) column sums (*i.e.*, sum of elements in a column) of a nonnegative matrix provide lower (resp. upper) bounds on its dominant eigenvalues. From (7.2.16) and (7.2.21), it follows that if we write

$$c_i = (\text{smallest column sum of } \boldsymbol{b}_i) \tag{7.3.3}$$

$$C_i = (\text{largest column sum of } \boldsymbol{b}_i) \quad i = 1, \ldots, r \tag{7.3.4}$$

then

$$\min_i c_i \leq \mu \leq \max_i C_i \tag{7.3.5}$$

$$\min_i c_i^2 \leq \mu \leq \max_i C_i^2.$$

The same argument used in equation (4.2.10) shows that

$$\sum_{i=1}^{r} \pi_i \log c_i \leq a \leq \sum_{i=1}^{r} \pi_i \log C_i, \tag{7.3.6}$$

where π_i = (Stationary probability of state i for the Markov chain with transition probability matrix $\boldsymbol{p}$).

Finally note that the matrix $\boldsymbol{d}$ from (7.2.8) satisfies (elementwise)

$$\boldsymbol{d} \geq \operatorname{diag}(\boldsymbol{d}_{11}, \boldsymbol{d}_{22}, \ldots, \boldsymbol{d}_{rr}). \tag{7.3.7}$$

Hence μ, the dominant eigenvalue of $\boldsymbol{d}$, will be an upper bound for the dominant eigenvalues of all the diagonal blocks on the right of (7.3.7),

$$\mu \geq \sup_i p_{ii} \lambda_i \tag{7.3.8}$$

where λ_i = (dominant eigenvalue of $\boldsymbol{b}_i$), $i = 1, \ldots, r$. Arguing similarly for $\boldsymbol{d}_2$ yields

$$\beta \geq \sup_i p_{ii} \lambda_i^2. \tag{7.3.9}$$

Cohen (1979a) provides interesting examples which show that $\boldsymbol{d}$ and $\boldsymbol{d}_2$ are not approximated by $\sum \pi_i \boldsymbol{b}_i$ and $\sum \pi_i (\boldsymbol{b}_i \otimes \boldsymbol{b}_i)$ respectively. The results above provide little qualitative insight into μ, β or a, and I now turn to a method which does.

4 Perturbation Expansions

The idea here is to ask what happens when the variability between environments (*i.e.*, the differences $b_i - b_j$) is small. It turns out in this case that one can get analytical expressions which describe accurately the behavior of μ and β (and, later, a). These analytical results yield considerable insight into the various growth rates.

The basic setup here appears again in later chapters. Define the **average projection matrix**

$$b = \sum_{i=1}^{r} \pi_i b_i = \mathsf{E}\, X_t \tag{7.4.1}$$

with the average taken over the stationary distribution of environmental states. Define also the deviations

$$h_i = b_i - b. \tag{7.4.2}$$

The strategy now is to decompose d (which determines μ) and d_2 (which determines β) into terms which depend on the average b, terms which are linear in the h_i, and terms quadratic in the h_i. Assuming that the hs are *small*, we then use perturbation techniques to find expansions for μ and β in powers of the hs. I will be very brief on the details here (see Tuljapurkar (1982b) for those) and focus on the results.

The first step is to rewrite d and d_2 from (7.2.16) and (7.2.21), using (7.4.1) and (7.4.2). We find

$$\begin{aligned} d &= d^{(0)} + d^{(1)}, \\ &= (p^T \otimes b) + \Delta(p^T \otimes I) \end{aligned} \tag{7.4.3}$$

with

$$\Delta = \operatorname{diag}(h_1, h_2, \ldots, h_r). \tag{7.4.4}$$

Similarly

$$\begin{aligned} d_2 &= d_2^{(0)} + d_2^{(1)} + d_2^{(2)} \\ &= (p^T \otimes b \otimes b) + (\Delta_1 + \tilde{\Delta}_1)(p^T \otimes I \otimes I) \\ &\quad + \Delta_2(p^T \otimes I \otimes I), \end{aligned} \tag{7.4.5}$$

with

$$\begin{aligned} \Delta_1 &= \operatorname{Diag}(b \otimes h_i), \\ \tilde{\Delta}_1 &= \operatorname{Diag}(h_i \otimes b), \\ \Delta_2 &= \operatorname{Diag}(h_i \otimes h_i). \end{aligned} \tag{7.4.6}$$

The superscripts in (7.4.3) and (7.4.6) indicate orders of h_i. For small hs, the dominant eigenvalues of d and d_2 can be computed as perturbations to the dominant eigenvalues of $\boldsymbol{d}^{(0)}$ and $\boldsymbol{d}_2^{(0)}$ respectively. I assume a full spectral decomposition for $\boldsymbol{b}$ and also for $\boldsymbol{p}$. The dominant eigenvalue of $\boldsymbol{b}$ is λ_0 (see Section 2.1), while that of $\boldsymbol{p}$ is 1; the dominant eigenvalues of $\boldsymbol{d}^{(0)}$ and $\boldsymbol{d}_2^{(0)}$ are just λ_0, λ_0^2 respectively.

Perturbation calculations following the usual methods of matrix analysis (Kato 1966) lead to the expansions

$$\mu \simeq \lambda_0 + \xi + \theta, \tag{7.4.7}$$
$$\beta \simeq \lambda_0^2 + \tau^2 + 4\lambda_0\xi + 2\lambda_0\theta, \tag{7.4.8}$$

accurate to second order in the $\boldsymbol{h}_i$. The terms in these expansions are

$$\tau^2 = (\boldsymbol{v}_0 \otimes \boldsymbol{v}_0)^T \boldsymbol{c}(0)(\boldsymbol{u}_0 \otimes \boldsymbol{u}_0), \tag{7.4.9}$$
$$\lambda_0\xi = (\boldsymbol{v}_0 \otimes \boldsymbol{v}_0)^T \boldsymbol{s}(\boldsymbol{u}_0 \otimes \boldsymbol{u}_0). \tag{7.4.10}$$

Here

$$\boldsymbol{c}(0) = \mathrm{Cov}[\boldsymbol{X}_t \otimes \boldsymbol{X}_t] \tag{7.4.11}$$

is the **one period variance-covariance** matrix of vital rates, while $\boldsymbol{S}$ is a sum of all **two-period covariances**,

$$\boldsymbol{s} = \sum_{\ell > 0} \boldsymbol{c}(\ell), \tag{7.4.12}$$
$$\boldsymbol{c}(\ell) = \mathrm{Cov}[\boldsymbol{X}_t \otimes \boldsymbol{X}_{t+\ell}], \quad \ell \geq 1.$$

The term θ is more complicated and requires the spectral decompositions of $\boldsymbol{p}$ and $\boldsymbol{b}$. Recall the matrix $\boldsymbol{q}$ of (2.2.10). Now let the eigenvalues and corresponding left and right eigenvectors of $\boldsymbol{p}$ be ν_m, $\boldsymbol{\psi}_m$, $\boldsymbol{\pi}_m$, $m = 0, 1, \ldots, r-1$. Of course $\nu_0 = 1$ and $\boldsymbol{\pi}_0 \equiv \boldsymbol{\pi}$, the vector of equilibrium probabilities for the Markov chain. Define the matrices

$$\boldsymbol{\alpha}_i = \sum_m \boldsymbol{\pi}_i(m)\boldsymbol{h}_m, \tag{7.4.13}$$
$$\boldsymbol{\delta}_i = \sum_m \boldsymbol{\pi}_0(m)\boldsymbol{\psi}_i(m)\boldsymbol{h}_m, \tag{7.4.14}$$

where $i, m = 0, 1, \ldots, r-1$. From (7.4.2) it follows that $\boldsymbol{\alpha}_0 = \boldsymbol{0}$, and because $\boldsymbol{\psi}_0 = \boldsymbol{e} = (11\ldots1)^T$, it also follows that $\boldsymbol{\delta}_0 = \boldsymbol{0}$. The term θ in (7.4.7) can now be written out as

$$\begin{aligned}
\lambda_0\theta = &\sum_{i>0} \nu_i \boldsymbol{v}_0^T \boldsymbol{\alpha}_i \boldsymbol{\delta}_i \boldsymbol{u}_0 \\
&- \sum_{i>0} \nu_i (\boldsymbol{v}_0 \otimes \boldsymbol{v}_0)^T (\boldsymbol{\alpha}_i \otimes \boldsymbol{\delta}_i)(\boldsymbol{u}_0 \otimes \boldsymbol{u}_0) \\
&+ \sum_{\substack{m>0 \\ i>0}} \nu_i^{m+1} \boldsymbol{v}_0^T \boldsymbol{\alpha}_i \boldsymbol{q}^m \boldsymbol{\delta}_i \boldsymbol{u}_0, \\
&i = 1, \ldots, r-1, \quad m = 1, 2, 3, \ldots
\end{aligned} \tag{7.4.15}$$

Important features of (7.4.7–8) are:

- With no variation $\mu = \lambda_0$, $\beta = \lambda_0^2$.
- With no serial autocorrelation $\big(i.e., \varsigma(\ell) = \mathbf{0}$ for $\ell > 1$ in (7.4.12)$\big)$ we have $\xi = \theta = 0$ and $\mu = \lambda_0$, $\beta = \lambda_0^2 + \tau^2$. This is the I.I.D. case.
- Serial correlation affects μ and β in very similar ways, while the one-period covariances *only* affect β.

Major qualitative conclusions we may draw from (7.4.7–8) are:

- β increases with increasing variance in every vital rate, and is increased (decreased) by positive (negative) pairwise covariances between vital rates.
- The dominant effects of autocorrelation are contained in ξ (see 7.4.10), while θ (see 7.4.14) depends relatively weakly on autocorrelation.
- If there is a single dominant autocorrelation (*e.g.*, in a 2–state Markov chain where there is a unique serial correlation, or in a multistate chain if one of the subdominant eigenvalues of $\boldsymbol{p}$ is much larger in magnitude than the rest), then both μ and β increase rapidly as this autocorrelation changes from -1 to $+1$. The effects of positive autocorrelation, especially near the limit, are much greater than of negative autocorrelation.
- Changes in environmental sequencing which affect neither the stationary frequencies of the states (*i.e.*, π_0) nor the correlation-determining eigenvalues ν_i of $\boldsymbol{p}$, *can* nevertheless change μ and β for the $r(>2)$ state process.

5 The Lognormal Approximation

The preceding results tell us about moments but not about a and σ^2. Recall the lognormal distribution of population size, (4.2.8); if we take this to be the actual distribution at large t, then

$$\mathsf{E}\, M_t \propto e^{at+\frac{1}{2}\sigma^2 t},$$
$$\mathsf{E}\, M_t^2 \propto e^{2at+2\sigma^2 t}.$$

This leads to the equations

$$a + \frac{1}{2}\sigma^2 \simeq \log\mu,$$
$$a + \sigma^2 \simeq \log\beta,$$

and thus

$$a \simeq 2\log\mu - \frac{1}{2}\log\beta, \tag{7.5.1}$$
$$\sigma^2 \simeq \log\beta - 2\log\mu. \tag{7.5.2}$$

This pair of equations was derived in Tuljapurkar (1982a); I follow Roerdink's (1987) terminology and call this the **lognormal approximation** to a and σ^2. This approximation has been found useful (Slade and Levenson 1982, Roerdink 1987) but is not well characterized. A different, more systematic approximation is presented later which uses direct perturbation analysis of the logarithmic moments (Tuljapurkar 1982b).

An interesting extension of this reasoning is to note that for any integer n one can compute the geometric growth rate $\mu(n)$, say, of $\mathsf{E}\, M_t^n$. Now the lognormality argument above shows that

$$\mathsf{E}\, M_t^n \propto \exp\left(nat + \frac{n^2}{2}\sigma^2 t\right).$$

Thus we conclude that

$$a = \frac{d}{dn} \log \mu(n)\bigg|_{n=0}.$$

This result is known in physics and mathematics, but seems difficult to apply here.

8

RANDOM SURVIVAL OR FERTILITY: EXACT RESULTS

A model which ought to be simple is one with 2 age classes and random survival or fertility. In this chapter we will see that this is only sometimes true depending on the question one asks. The examples should breathe some life into the general theory.

1 Random Survival Rate

The model I study is

$$\begin{pmatrix} N_{t+1}(1) \\ N_{t+1}(2) \end{pmatrix} = \begin{pmatrix} m_1 & m_2 \\ P_{t+1} & 0 \end{pmatrix} \begin{pmatrix} N_t(1) \\ N_t(2) \end{pmatrix}, \tag{8.1.1}$$

with m_1, $m_2 > 0$ and P_t being a time-varying random survival rate with $0 < p_1 \leq P_t \leq p_2 \leq 1$. It is essential to convert (8.1.1) into an equation for age structure, because with 2 age classes the age structure is described by one number, *e.g.*, the proportion in age class 1. An alternative and more useful variable describing age structure is the ratio of old to young,

$$U_t = N_t(2)/N_t(1). \tag{8.1.2}$$

I will first show that if P_t is bounded, then so is U_t; second, I obtain the exact probability distribution of U_t when P_t is I.I.D. and takes on only 2 possible values; third, I obtain the exact distribution of U_t when P_t follows a 2–state Markov chain.

1.1 Bounds on age structure

The ratio of old to young obeys the equation

$$U_{t+1} = \frac{P_{t+1}}{m_1 + m_2 U_t} = f(P_{t+1}, U_t). \tag{8.1.3}$$

Note that $f(\cdot, u)$ is convex decreasing in u and that $f(p_1, u) \leq f(P_t, u) \leq f(p_2, u)$ for every u. Write $f_i(u) = f(p_i, u)$, $i = 1, 2$.

Equation (8.1.1) will show, irrespective of how the P_ts are chosen, that U_t is eventually bounded. Start with any u_0 and observe that

$$0 \leq U_1 = f(P_1, u_0) \leq f_2(u_0) \leq f_2(0);$$

at the next iteration,

$$U_2 = f(P_2, U_1) \le f_2(0),$$

while

$$U_2 = f(P_2, U_1) \ge f(P_2, f_2(0)) \ge f_1(f_2(0)) = f_1 \circ f_2(0).$$

Continuing in this fashion gives a sequence of bounds

$$\ell_n \le U_n \le m_n,$$

with

$$\begin{array}{rclcl} m_n & = & f_2(\ell_{n-1}) & = & f_2 \circ f_1(m_{n-2}), \\ \ell_n & = & f_1(m_{n-1}) & = & f_1 \circ f_2(\ell_{n-1}). \end{array} \tag{8.1.4}$$

The functions in (8.1.4) have simple properties for the following reason. Take any population vector with ratio u of old to young, apply first the Leslie matrix with survival rate p_2, then the Leslie matrix with survival rate p_1; the new population vector you get has an old to young ratio of $f_1 \circ f_2(u)$. Thus the behavior of iterates of $f_1 \circ f_2$ is given by the behavior of powers of the product of these two Leslie matrices. So in (8.1.4) one is assured that as n increases to infinity $m_n \to \beta$ and $\ell_n \to \alpha$ where

$$\begin{array}{rclcl} \beta & = & f_2(\alpha) & = & f_2 \circ f_1(\beta), \\ \alpha & = & f_1(\beta) & = & f_1 \circ f_2(\alpha). \end{array} \tag{8.1.5}$$

Since the limit of large n for any initial u_0 is the stochastic steady state for U_n, one has

$$\alpha \le U_t \le \beta \tag{8.1.6}$$

in the steady state.

It should be clear that the bounds (8.1.6) apply so long as $p_1 \le P_t \le p_2$, irrespective of the actual distribution of survival rate within these bounds. Also from (8.1.1) note that bounds similar to (8.1.6) may be computed for variation in other vital rates. These bounds are an example of the general bounds obtained in Section 9.1.

1.2 I.I.D. Survival Rate

This section presents the derivation of an exact, explicit distribution of age-structure for (8.1.1) in the case when survival P_t takes one of two values with fixed probability. While this might seem the ideal special case, the results reveal unexpected complexity.

A. *Equation for distribution of age structure*

The survival rate P_t in (8.1.1) is taken to be a sequence of independent identically distributed random variables bounded as in the previous section. Suppose that the probability density function of P_t is $k(\cdot)$, and write the inverse function to $f(p, u)$ in (8.1.3) as $g_p(u)$. The steady state distribution of U_t has the property of reproducing itself under the action of (8.1.3). Thus

$$\begin{aligned} V(x) &= \Pr\{U_{t-1} \text{ in } (0, x]\} \\ &= \int dw\, k(w) \Pr\{U_t \text{ in } g_w((0, x])\} \\ &= \int dw\, k(w) \Pr\{U_t \text{ in } [g_w(x), \infty)\} \\ &= 1 - \int dw\, k(w) V(g_w(x)), \end{aligned} \tag{8.1.7}$$

where $\Pr\{A\}$ = Probability of A. Equation (8.1.7) has long been known in the physics literature as the Dyson-Schmidt equation (Dyson 1953, Schmidt 1957); a demographic version is given by Cohen (1977a,b).

Here I focus on dichotomous random survival with $k(w) = \pi\delta(w - p_1) + (1 - \pi)\delta(w - p_2)$. Then (8.1.7) reduces to the functional equation

$$V(x) = 1 - \pi V(g_1(x)) - (1 - \pi) V(g_2(x)), \tag{8.1.8}$$

where the shorthand $g_1 \equiv g_{p_1}$ is used. From (8.1.6) one adds the information

$$V(x) = \begin{cases} 0, & x < \alpha, \\ 1, & x \geq \beta. \end{cases} \tag{8.1.9}$$

B. *The singular solution*

The solution to (8.1.8) is now constructed (the following discussion is based on Tuljapurkar 1984a). From (8.1.9) note that

$$V(g_i(x)) = \begin{cases} 0, & g_i(x) < \alpha, \\ 1, & g_i(x) \geq \beta. \end{cases} \tag{8.1.10}$$

for $i = 1, 2$. Now $g_i(x) < \alpha$ is the same as $x > f_i(\alpha)$, and thus

$$\begin{aligned} V(g_1(x)) = 0 \quad &\text{for} \quad x > f_1(\alpha) = \alpha_1, \\ V(g_2(x)) = 1 \quad &\text{for} \quad x < f_2(\beta) = \beta_1. \end{aligned} \tag{8.1.11}$$

Looking back at (8.1.8) observe that *if*

$$f_1(\alpha) = \alpha_1 < \beta_1 = f_2(\beta), \tag{8.1.12}$$

then in the interval $\alpha_1 < x < \beta_1$ one can use (8.1.9) in (8.1.8) and get $V(x) = \pi$.

The above argument may now be repeated. One has $V(g_1) = \pi$ for $\alpha_1 < g_i < \beta_1$ and thus $V(g_1(x)) = \pi$ for $f_1(\beta_1) < x < f_1(\alpha_1)$, while $V(g_2(x)) = \pi$ for $f_2(\beta_1) < x < f_2(\alpha_1)$. Using (8.1.8) and (8.1.9) yields $V(x) = 1 - \pi(\pi) - (1-\pi)(1) = \pi(1-\pi)$ for $f_1(\beta_1) < x < f_1(\alpha_1)$, and $V(x) = 1 - \pi(1-\pi)$ for $f_2(\beta_1) < x < f_2(\alpha_1)$.

A general pattern can now be discerned. The first step ($m = 1$) yields $2^{m-1} = 1$ interval, *i.e.*, $\alpha_1 < x < \beta_1$ in which $V(x) = \pi = V_{11}$. The next step ($m = 2$) yields $2^{m-1} = 2$ intervals, $f_1(\beta_1) < x < f_1(\alpha_1)$ with $V = V_{21} = \pi(1-\pi)$, and $f_2(\beta_1) < x < f_2(\alpha_1)$ with $V = V_{22} = 1 - \pi(1-\pi)$. At the mth stage there are 2^{m-1} intervals in which V takes on values V_{ms} for $s = 1, \dots, 2^{m-1}$. To iterate the procedure algebraically is straightforward but tedious. For illustration the next iteration step yields (proceeding from left to right on (α, β)):

$$
\begin{array}{ll}
V_{21} = \pi^2(1-\pi), & f_1 \circ f_2(\alpha_1) < x < f_1 \circ f_2(\beta_1), \\
V_{22} = \pi[1 - \pi(1-\pi)], & f_1 \circ f_1(\alpha_1) < x < f_1 \circ f_1(\beta_1), \\
V_{23} = \pi[1 + (1-\pi)^2], & f_2 \circ f_2(\alpha_1) < x < f_2 \circ f_2(\beta_1), \\
V_{24} = 1 - \pi(1-\pi)^2, & f_2 \circ f_1(\alpha_1) < x < f_2 \circ f_1(\beta_1),
\end{array}
$$

Let the sth interval at the mth step be $(d_{ms}\ t_{ms})$. The steplike singular character of the distribution may be summarized in the equation

$$
V(x) = \sum_{m=1}^{\infty} \sum_{s=1}^{2^{m-1}} V_{ms}[\theta(x - t_{ms}) - \theta(x - d_{ms})] + \theta(x - \beta). \qquad (8.1.13)
$$

Here $\theta(x)$ is the unit step function and the V_{ms} are as described earlier. From (8.1.13) a density can be formally written as an infinite sum of delta functions.

Figures 8.1.1 and 8.1.2 illustrate the nature of $V(x)$. Note the pattern of irregularity. If one plots this function at a higher resolution, the irregularity deepens as more steps appear. This scale-dependent character of $V(x)$ is reminiscent of Cantor functions.

The only difference between Figures 8.1.1 and 8.1.2 is the value of π. Although the interval structure which defines the steplike nature of $V(x)$ is independent of π, the weight in these intervals changes sharply with π as shown.

As a final note, it is useful to be able to compute expectations of functions (*e.g.*, growth rate) of the type $G(U_t, P_{t+1})$. In the steady state using (8.1.13) one has

$$
\begin{aligned}
\mathsf{E}\, G(U_t, P_{t+1}) = \pi G(\beta, p_1) + \pi \sum_{m=1}^{\infty} \sum_{s=1}^{2^{m-1}} V_{ms}\left[G(t_{ms}, p_1) - G(d_{ms}, p_1)\right] \\
+ \{\text{terms identical to these but with } p_2 \text{ replacing } p_2 \\
\text{and } (1-\pi) \text{ replacing } \pi\}.
\end{aligned}
$$

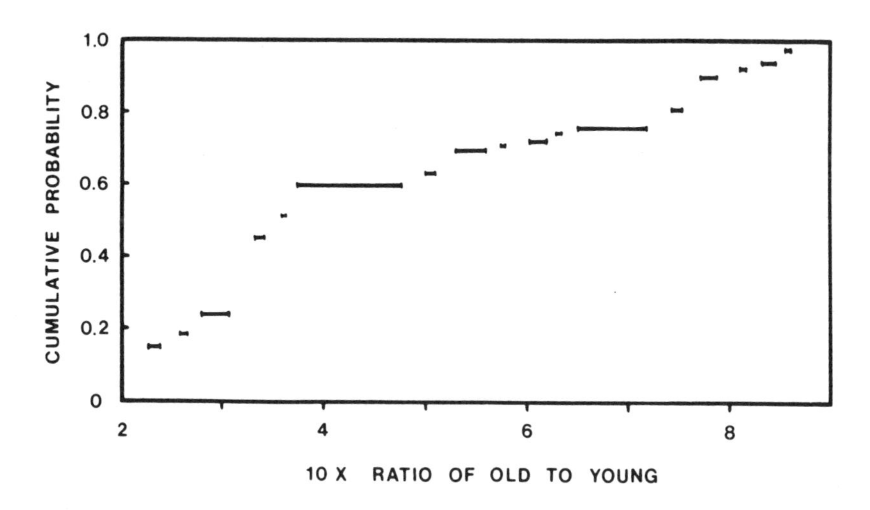

FIGURE 8.1.1. Exact distribution V, above threshold, with $p_1 = 0.3$, $p_2 = 0.7$, $m_1 = 1$, $\pi = 0.3$

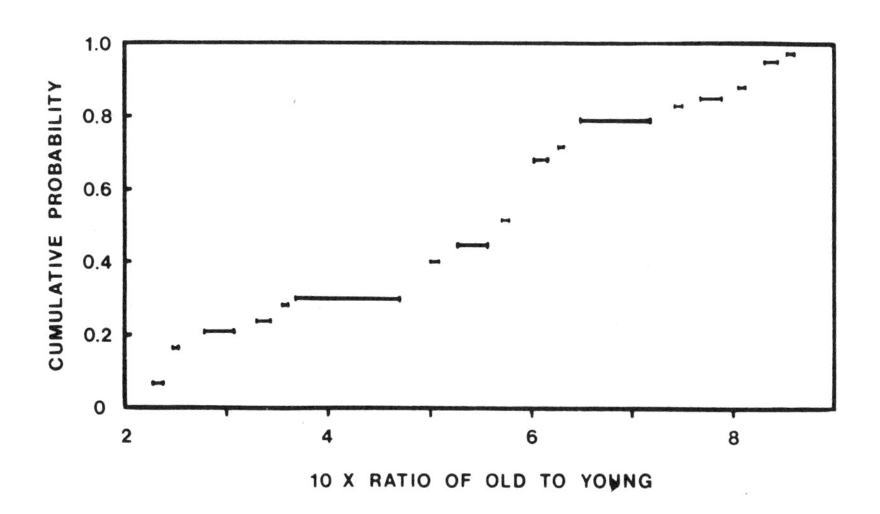

FIGURE 8.1.2. Exact distribution V, above threshold, with $\pi = 0.6$ but other parameters as in preceding figure

From the nature of the intervals (d_{ms}, t_{ms}) one knows that the lengths of these intervals decrease geometrically fast with m, so one can hope that the series above will converge rapidly.

C. *The transition*

The steplike solution (8.1.13) hinges on the condition (8.1.12) that $\alpha_1 < \beta_1$. This inequality can be rewritten, using the explicit function (8.1.3) and the definitions $p_1 = p - \delta$, $p_2 = p + \delta$, as

$$m_2 p(\beta + \alpha) - m_2 \delta(\beta - \alpha) < 2m_1 \delta.$$

Now use the left-most equations in (8.1.5) to get

$$\beta - \alpha = 2\delta / m_1.$$

The other equalities in (8.1.5) yield separate quadratic equations for α and β respectively. Adding together the solutions of these equations one gets

$$\beta + \alpha = \frac{m_1}{m_2} + \left(\frac{m_1^2}{m_2^2} + \frac{4p}{m_2} + \frac{4^2}{m_1^2} \right)^{1/2}.$$

Insert the two expressions above into the inequality, square and simplify to find that $\alpha_1 < \beta_1$ when

$$\delta^2 > (1 - 2z)p^2 \tag{8.1.14}$$

where $z = (m_1^2 / m_2 p)$. For z greater than $\frac{1}{2}$, this inequality holds for all nonzero δ. As z decreases below 0.5, it passes through the threshold value $(1 - \delta^2/p^2)/2$ below which α_1 exceeds β_1 and the interval structure exploited earlier disappears. The construction of (8.1.12) then no longer works.

However, a numerical solution of the functional equation (8.1.8) is easily implemented and shows that below the threshold z the distribution $V(x)$ appears absolutely continuous. See for an example Figure 8.1.3 in which I show $V(x)$ below threshold; compare this with Figure 8.1.1 above threshold.

Although I have no analytical solution for $V(x)$ below threshold, it is possible to obtain information on the behavior of $V(x)$ near the limits as $x \to \alpha$ and $x \to \beta$. To do this it is helpful to iterate the functional equation (8.1.8) once and get

$$\begin{aligned} V(x) = \pi^2 V(g_{11}(x)) + \pi(1 - \pi)V(g_{12}(x)) \\ +\pi(1 - \pi)V(g_{21}(x)) + (1 - \pi)^2 V(g_{11}(x)), \end{aligned}$$

where I use the notation $g_{ij}(x) \equiv g_i \circ g_j(x)$. Notice that as $x \to \alpha$ from above, $g_{11}(x)$, $g_{12}(x)$ and $g_{22}(x)$ are all eventually $< \alpha$, so for x close enough to α the equation above reduces to (recall(8.1.10)),

$$V(x) = \pi(1 - \pi)V(g_{21}(x)).$$

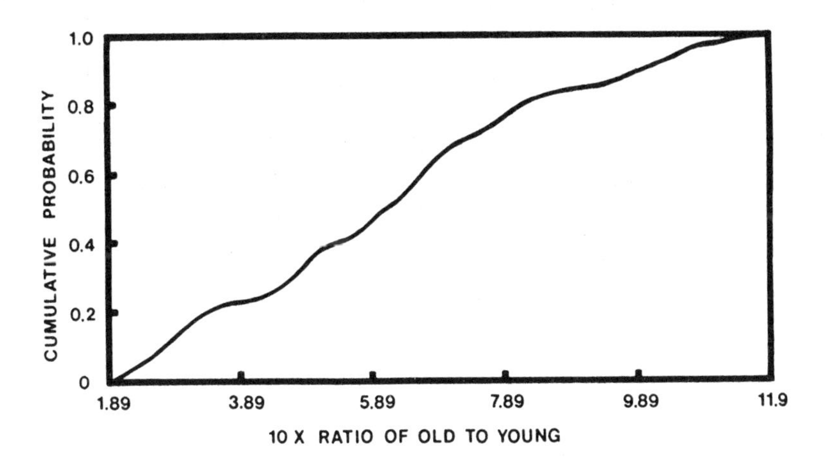

FIGURE 8.1.3. Numerically computed distribution V, below threshold, with $m_1 = 0.4$, $p_1 = 0.3$, $p_2 = 0.7$, $\pi = 0.3$.

Now α is a fixed point of g_{21} and so $g_{21}(x)$ can be written as a product $(x - \alpha)b_{21}(x)$, where $b_{21}(x)$ is positive for $x \geq \alpha$. With this information one can verify as $x \to \alpha$ the equation for V is satisfied by

$$V(x) \sim c\,(x - \alpha)^{\theta}$$

where $\theta = (\text{constant})|\log \pi(1 - \pi)|$. The quantity c and the constant in θ are numbers which do not depend on π. A similar form holds as $x \to \beta$,

$$V(x) \sim 1 - c'(\beta - x)^{\theta'},$$

$$\theta' = (\text{constant})'|\log \pi(1 - \pi)|.$$

The piling up of probability mass as $\pi \to 0$ or 1 is intuitively reasonable.

The feature of a threshold separating singular and continuous distributions is intrinsic to the dichotomous nature of the distribution of survival rate. Consequently one expects that a more complex but dichotomous stochastic process for P_t will not change this feature; in the next section I show this to be so.

1.3 Serially correlated survival

A. *Equation for distribution*

The survival rate P_t in (8.1.1) is here assumed to take values $p_1 = p - \delta$, $p_2 = p + \delta$, according as a Markov process is in state 1, 2 respectively at time t. The 2–state Markov process has a transition probability matrix

$$\boldsymbol{Q} = \begin{pmatrix} q_{11} & q_{12} \\ q_{21} & q_{22} \end{pmatrix} = \begin{pmatrix} \pi + \rho(1 - \pi) & (1 - \pi)(1 - \rho) \\ \pi(1 - \rho) & 1 - \pi + \pi\rho, \end{pmatrix} \qquad (8.1.15)$$

where π = (stationary probability of state 1) and ρ is the serial autocorrelation. In my notation $q_{ij} = \Pr\{\text{Going from state } i \text{ to state } j\}$, $i, j = 1, 2$. The joint process (U_t, P_t) is now Markovian and this complicates matters a little.

As before, the inverse function to $f(p, u)$ in (1) is $g_p(u)$ and the subscripts 1, 2 refer to values p_1, p_2 respectively. The joint distribution of U_t, P_t can be described by the functions

$$V_i(x) = \Pr\{U_t \le x \quad \text{and} \quad P_t = p_i\}, \qquad i = 1, 2.$$

A slight generalization of the arguments leading to (8.1.7) gives the pair of equations

$$V_i(x) = q_{1i}\left[V_1(\infty) - V_1\left(g_i(x)\right)\right] + q_{2i}\left[V_2(\infty) - V_2\left(g_i(x)\right)\right], \quad i = 1, 2.$$

These equations are essentially those of Schmidt (1957) and Cohen (1977b). To proceed recall that joint and conditional probabilities are related by

$$V_i(x) = \Pr\{U_t \le x | P_t = p_i\} \Pr\{P_t = p_t\}, \qquad i = 1, 2$$

with $\Pr\{A|B\} = \Pr\{A \text{ conditional on } B\}$. Also note that the bounds of Section 8.1.1 are still good so that $\Pr\{U_t \le \beta | P_t = p_i\} = 1$ for $i = 1, 2$. Thus

$$\begin{aligned} V_1(\infty) &= V_1(\beta) = \pi, \\ V_2(\infty) &= V_2(\beta) = 1 - \pi, \\ V_1(\alpha) &= V_2(\alpha) = 0. \end{aligned} \tag{8.1.16}$$

Using these values in the earlier equations, together with the fact that $(\pi, 1 - \pi)Q = (\pi, 1 - \pi)$, gives the functional equation

$$\begin{aligned} V_1(x) &= \pi - q_{11} V_1\left(g_1(x)\right) - q_{21} V_2\left(g_1(x)\right), \\ V_2(x) &= 1 - \pi - q_{12} V_1\left(g_2(x)\right) - q_{22} V_2\left(g_2(x)\right). \end{aligned} \tag{8.1.17}$$

The pair (8.1.17) subject to (8.1.16) replaces (8.1.8) and (8.1.9) of the previous section.

B. *The singular solution*

Equation (8.1.17) is solved in precisely the same fashion as (8.1.8). So long as the value of z is above threshold (*cf.* discussion around (8.1.14)) one has condition (8.1.9). Begin by noting that

$$\begin{aligned} &V_1\left(g_1(x)\right) = V_2\left(g_1(x)\right) = 0 \qquad \text{for } x > \alpha_1, \\ &\left.\begin{aligned} V_1\left(g_2(x)\right) &= \pi \\ V_2\left(g_2(x)\right) &= 1 - \pi \end{aligned}\right\} \qquad \text{for } x < \beta_1. \end{aligned} \tag{8.1.18}$$

Proceed in steps as in the previous section. In the first step ($m = 1$) use (8.1.18) in (8.1.17) to conclude that

$$\left.\begin{array}{rcl} V_1(x) & = & \pi, \\ V_2(x) & = & 0, \end{array}\right\} \quad \text{for } \alpha_1 < x < \beta_1 \tag{8.1.19a}$$

At the next ($m = 2$) step use (8.1.19a) and (8.1.18) in (8.1.17) to get

$$\left.\begin{array}{rcl} V_1(x) & = & \pi(1-\pi)(1-p), \\ V_2(x) & = & 0, \end{array}\right\} \text{for } f_1(\beta_1) < x < f_1(\alpha_1),$$

$$\left.\begin{array}{rcl} V_1(x) & = & \pi \\ V_2(x) & = & (1-\pi)[1-\pi(1-\rho)], \end{array}\right\} \text{for } f_2(\beta_1) < x < f_2(\alpha_1). \tag{8.1.19b}$$

The next ($m = 3$) step is the last one I will report:

$$\left.\begin{array}{l} V_1(x) = \pi^2(1-\pi)(1-\rho^2) \\ V_2(x) = 0 \end{array}\right\} \quad \text{for } f_1 \circ f_2(\alpha_1) < x < f_1 \circ f_2(\beta_1),$$

$$\left.\begin{array}{l} V_1(x) = \pi - \pi^2(1-\pi)(1-\rho) \\ \qquad\qquad + \pi(1-\pi)^2\rho(1-\rho) \\ V_2(x) = 0 \end{array}\right\} \quad \text{for } f_1 \circ f_1(\alpha_1) < x < f_1 \circ f_1(\beta_1),$$

$$\left.\begin{array}{l} V_1(x) = \pi \\ V_2(x) = \pi(1-\pi)(1-\rho)[1-\pi(1-\rho)] \end{array}\right\} \quad \text{for } f_2 \circ f_2(\alpha_1) < x < f_2 \circ f_2(\beta_1),$$

$$\left.\begin{array}{l} V_1(x) = \pi \\ V_2(x) = (1-\pi)[1-\pi(1-\pi)(1-\rho)^2] \end{array}\right\} \quad \text{for } f_2 \circ f_1(\alpha_1) < x < f_2 \circ f_1(\beta_1). \tag{8.1.19c}$$

The preceding values are written in such a way that the effects of changing ρ can be quickly assessed for fixed π. In particular, note that the unconditional probability distribution of U_t is given by

$$V(x) = \Pr\{U_t \le x\} = V_1(x) + V_2(x).$$

Thus in each interval a handy check on the results is to compute $V(x)$, put $\rho = 0$, and see if the results of Section 8.1.2 are recovered. As you may verify, they are.

Figures 8.1.4 and 8.1.5 illustrate the effect of changing autocorrelation on V_1, V_2 and V. These may be compared with Figure 8.1.1 which is drawn for the same π value and of course has $\rho = 0$. Note that positive autocorrelation tends to concentrate the distribution and make it appear less irregular.

The expectation of a function $G(U_t, P_{t+1})$ can easily be written out. Using the same notation for intervals as in (8.1.13) let $V_{ms}(i)$ be the value

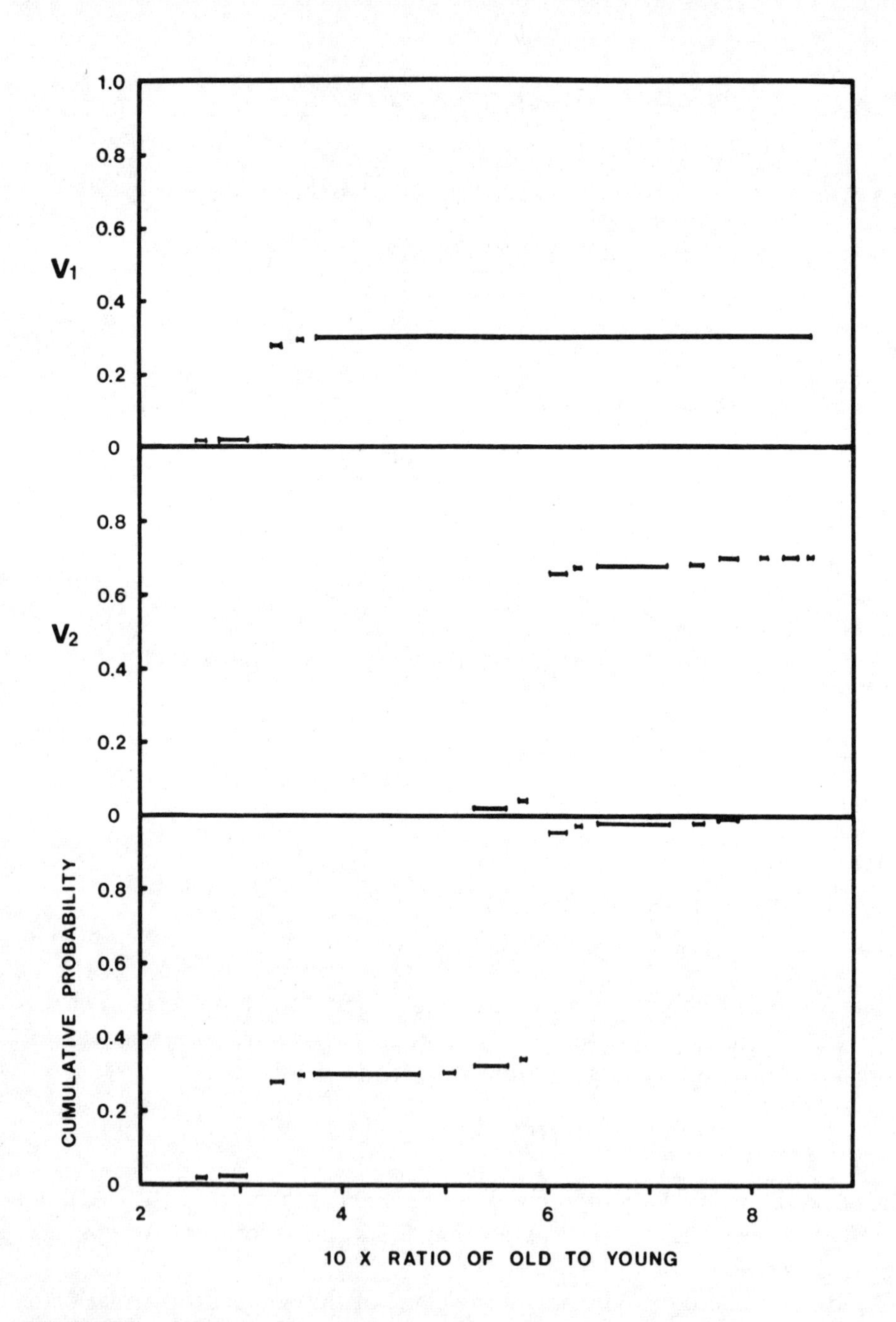

FIGURE 8.1.4. Panels show exact distribution; from the top, V_1, V_2, unconditional distribution $V_1 + V_2$. Serial autocorrelation $\rho = 0.9$, $p_1 = 0.3$, $p_2 = 0.7$, $m_1 = 1$, $\pi = 0.3$

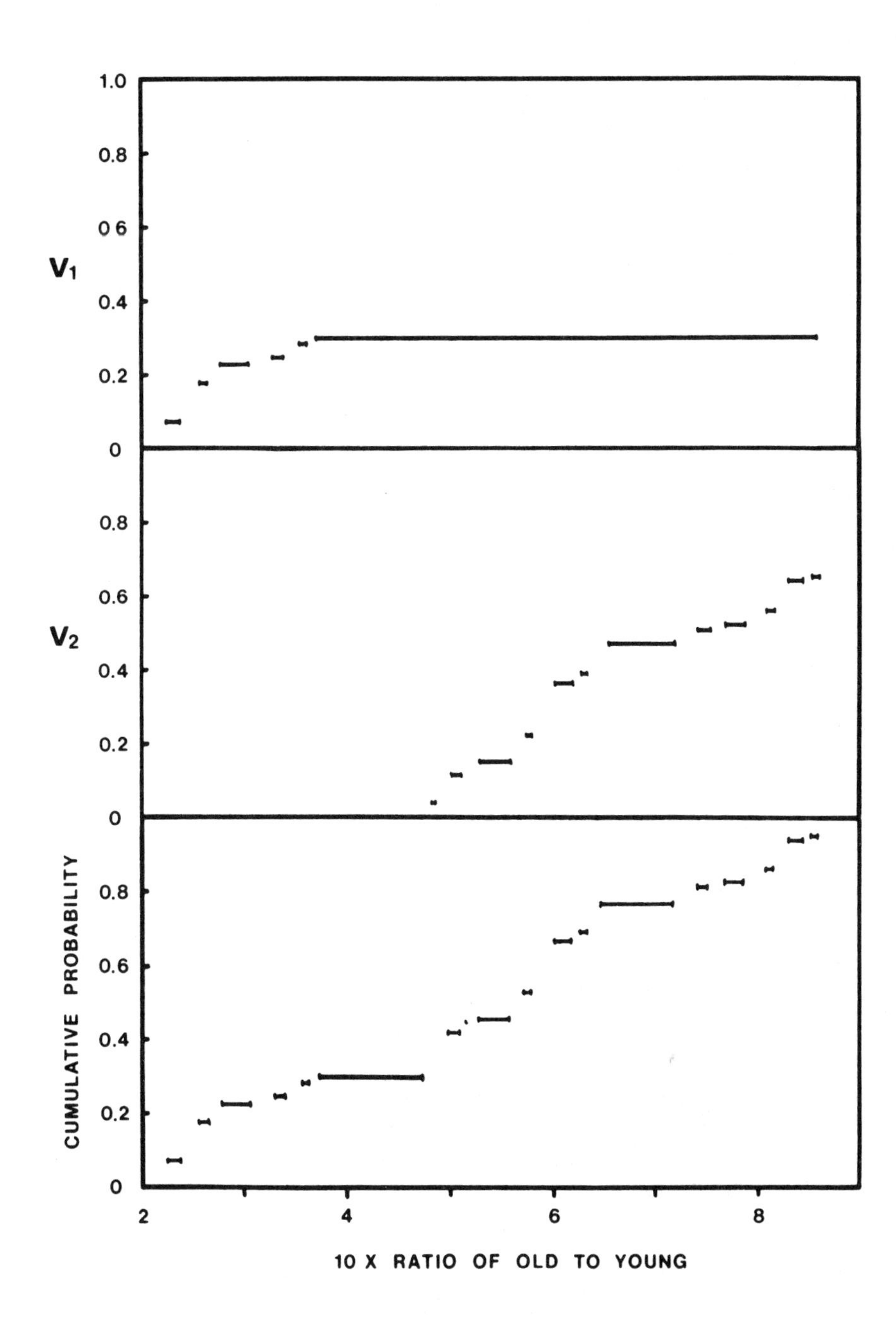

FIGURE 8.1.5. Exact distribution, parameters as in preceding figure except that here there is negative serial autocorrelation $\rho = -0.1$

of $V_i(x)$ in the sth interval at the mth step. Then

$$\begin{aligned} \mathsf{E}G(U_t, P_{t+1}) &= \sum_{i,j=1}^{2} q_{ij} \int dV_i(x) G(x, p_i) \\ &= \pi G(\beta, p_1) + (1-\pi) G(\beta, p_2) \\ &\quad + \sum_{i,j=1}^{2} q_{ij} \sum_{m=1}^{\infty} \sum_{s=1}^{2^{m-1}} V_{ms}(i) \left[G(t_{ms,p_j}) - G(d_{ms}, p_j) \right]. \end{aligned}$$

C. *The threshold*

Since the interval structure which produces the distribution (8.1.19) is the same as produced the distribution of Section 8.1.2, one expects identical threshold behavior. Here I have no analytical results but numerical work suggests that the distribution in the Markovian case is also smoothed below threshold. See Figure 8.1.6 for an example which should be compared with Figures 8.1.4 and 8.1.5.

As in Section 8.1.2, one can extract information on the asymptotic behavior of $V_i(x)$ as x tends to α or β. A similar but slightly more complicated procedure yields the present case the result that as $x \to \alpha$,

$$\begin{aligned} & V_1(x) \sim c''(x-\alpha)^{\theta'}, \\ & V_2(x) = 0, \\ & \qquad \theta'' = (\text{constant})'' |\log \pi(1-\pi)(1-\rho)|. \end{aligned}$$

The singularity which occurs in Section 8.1.2 as $\pi \to 0$ is here given added dimension by the autocorrelation factor $(1-\rho)$. As one might expect the limit of strong positive autocorrelation pulls the distribution towards its endpoints. An analog of the above holds in the limit as $x \to \beta$.

1.4 The Lessons of this Example

The feature that stands out is the complexity of the distribution for such a "simple" model. It is unpleasant to discover that while we have the distribution, it is difficult to compute objects like a from sums such as the one following (8.1.13). Numerical experiments suggest that it is difficult to sum over the Cantor set structure of (8.1.13), and that recursive computations don't converge rapidly. On the other hand, it is nice to see explicitly the smearing-out of age structure.

It should be clear that the method generalizes. The reader may, for example, apply these techniques to Cohen's (1979a) examples of 3-state Markov environments. Another extension is to a dichotomous I.I.D. survival rate in a 3 age class model; there the resulting 2-dimensional distribution of age structures will be supported by a Sierpinski gasket (or carpet). (Who dat? See Mandelbrot 1982).

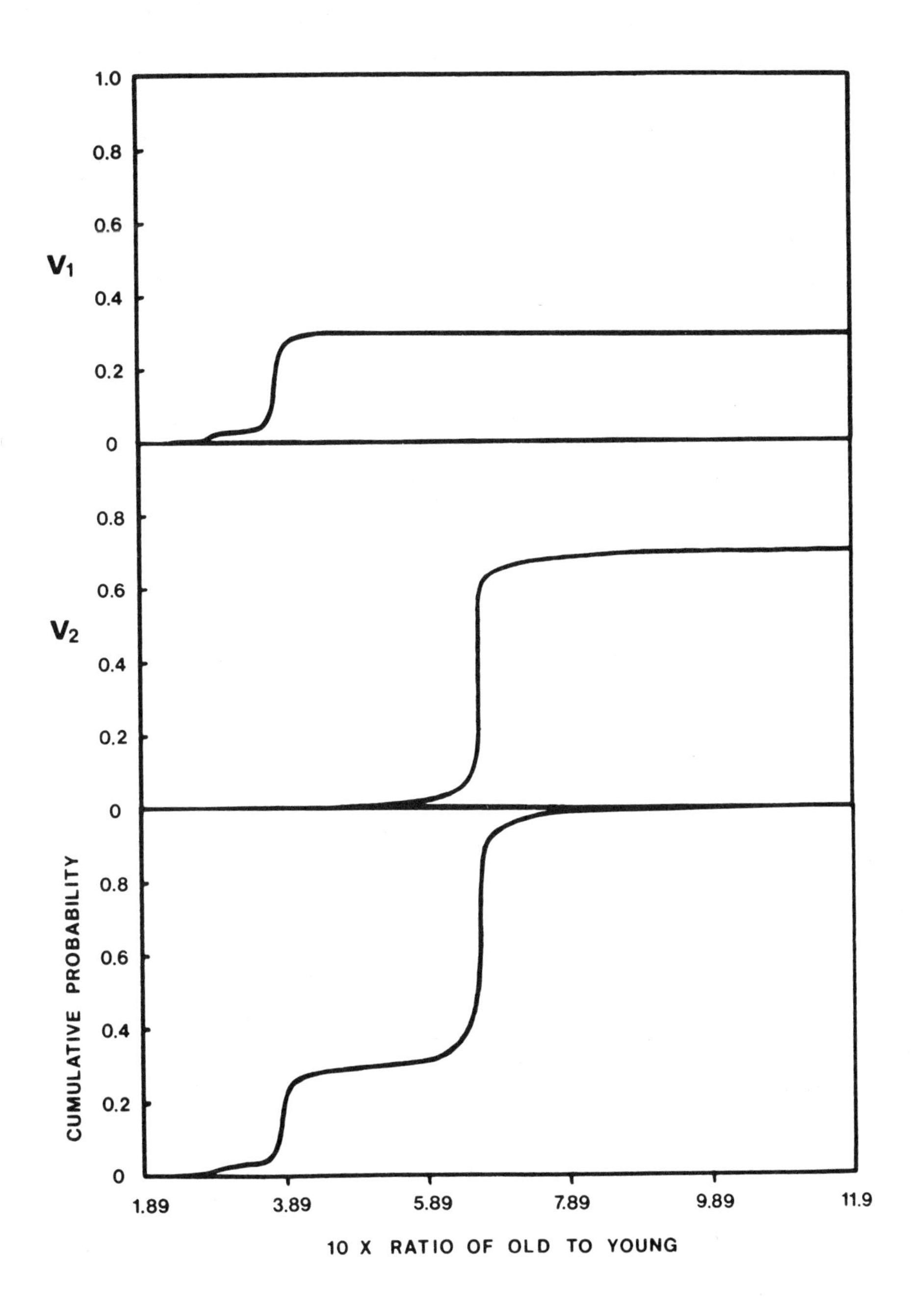

FIGURE 8.1.6. Numerically computed distributions below threshold, with $m_1 = 0.4$, $p_1 = 0.3$, $p_2 = 0.7$, $\pi = 0.3$, $\rho = 0.9$

A final extension is to deal with random fertilities in (8.1.1) instead of survival rates. As long as only one fertility varies, or both fertilities vary together, results precisely analogous to those of this section are easily obtained.

General conditions for the age structure to have a smooth or a singular distribution are not known. The state of understanding appears to reflect Hasminskii's (1980, p. 220) comment that the solution of equations such as (8.1.7) is "fraught with difficulties."

2 Random Fertility

I turn now to a 2 age class model for which I obtain an exact *and* analytic distribution of age structure. The results have been applied to the interesting biological question of why delayed flowering might have evolved in biennial plants and to the evolution of diapause and other prereproductive delays (see Chapter 16). The model is

$$\boldsymbol{N}_{t+1} = \boldsymbol{X}_{t+1}\boldsymbol{N}_t, \tag{8.2.1}$$

with

$$\boldsymbol{X}_t = \begin{pmatrix} m_1 F_t & m_2 F_t \\ p & 0 \end{pmatrix}. \tag{8.2.2}$$

Here $m_1 > 0$, $m_2 > 0$, $1 > p > 0$ and $\{F_t,\ t = 0, 1, \ldots\}$ is a sequence of I.I.D. random variables. I suppose that $(1/F_t)$ has a gamma distribution with probability density function

$$g(w) = (n^n/(n-1)!)\, w^{n-1} e^{-nw}. \tag{8.2.3}$$

The mean and variance of F_t are

$$\begin{aligned} \mathsf{E}\, F = \langle F \rangle &= \int_0^\infty dw\, g(w)(1/w) \\ &= n/(n-1) \sim 1 + (1/n) \end{aligned} \tag{8.2.4}$$

$$\begin{aligned} \text{Var}\ F = &\ \text{Variance}(F) = \mathsf{E}\, F^2 - \langle F \rangle^2 \\ &= \left\{(n-2)(1-1/n)^2\right\}^{-1} \sim (1/n) \end{aligned} \tag{8.2.5}$$

where the asymptotic limits are for large n. The parameter n controls the variance of fertility and when $n \to \infty$ we get $F_t \to 1$.

The **average projection matrix** from (8.2.2) is

$$\boldsymbol{b} = \begin{pmatrix} m_1\langle F \rangle & m_2\langle F \rangle \\ p & 0 \end{pmatrix} \tag{8.2.6}$$

with dominant eigenvalue λ_0 solving

$$\lambda_0^2 - m_1\langle F\rangle\lambda_0 - m_2 p\langle F\rangle = 0. \tag{8.2.7}$$

For this average matrix, the stable ratio of old to young is $\hat{u} = (p/\lambda_0)$ and so we can write

$$r_0 = \log\lambda_0 = \log\,(p/\hat{u}). \tag{8.2.8}$$

Also the convergence rate of a population governed by $\boldsymbol{b}$ to the stable state is determined by the subdominant eigenvalue λ_1 of $\boldsymbol{b}$, and we have

$$r_1 = \log|\lambda_1| = \log\,(m_2\langle F\rangle p) - r_0. \tag{8.2.9}$$

These two equations will be useful for comparison with the stochastic results to follow.

2.1 Equation for distribution

The equation for U_t, the ratio of old to young in (8.2.1), is

$$U_{t+1} = \frac{p}{F_{t+1}(m_1 + m_2 U_t)}. \tag{8.2.10}$$

It will be convenient to work with the quantity

$$R_t = (m_2 U_t/m_1), \tag{8.2.11}$$

so that

$$R_{t+1} = \frac{1}{zF_{t+1}}\cdot\frac{1}{(1+R_t)}, \tag{8.2.12}$$

with $z = (m_1^2/m_2 p)$. In this model we have $0 < F_t < \infty$ and so $0 < R_t < \infty$. Assuming that in the steady state R_t has a probability density function $C(x)$, and recalling that (8.2.3) is the density of $(1/F_t)$, we can write

$$C(x) = \int_0^\infty dy\, C(y)\int_0^\infty dw\,\delta\left[x - \frac{w}{z(1+y)}\right] g(w),$$

where $\delta(\cdot)$ is the Dirac delta function. Changing variables to do the integral over w produces an equation for C:

$$C(x) = \int_0^\infty dy\, C(y) zx(1+y) g\left[zx(1+y)\right]. \tag{8.2.13}$$

2.2 Distribution

Inserting the explicit (8.2.3) into (8.2.13) yields an integral equation which was (happily) solved by Dyson (1953), and the solution (verifiable by substitution) is

$$C(x) = \kappa^{-1}x^{n-1}(1+x)^{-n}e^{-nzx}, \tag{8.2.14}$$

where κ is a constant which ensures that the area under $C(x)$ is unity,

$$\kappa = \int_0^\infty dx\, x^{n-1}(1+x)^{-n}e^{-nzx} \tag{8.2.15}$$

Figure 8.2.1 plots C for three different values of n. For $n = 10$ and a coefficient of variation in fertility of 0.3, the distribution of age structure is seen to have substantial variance with a peak quite different from the "stable" value determined by the average vital rates in $\boldsymbol{b}$.

The smoothness of C allows us to compute expectations easily, in contrast to Section 8.1. Figure 8.2.2 plots the average $\mathsf{E}\, R_t$ as a function of $(1/n)$; Figure 8.2.3 plots the variance of R_t as n changes.

2.3 Growth Rate

We can compute a using (4.2.10) which translates here into a double integral. With $M_t =$ (population at time t), (8.2.11) and (8.2.1) show that

$$\begin{aligned} a = \mathsf{E} \log(M_{t+1}/M_t) &= \mathsf{E} \log[m_1 F_t(1+R_t)+p] - \mathsf{E} \log[m_2 + m_1 R_t] + \log m_2 \\ &= \int dx \int dw C(x) g(w) \log\left[\frac{m_1}{w}(1+x)+p\right] \\ &\quad - \int dx C(x) \log(m_2 + m_1 x) + \log m_2. \end{aligned}$$

However, it is easier to use (8.2.12) in the first line of the above equation and get

$$a = \mathsf{E} \log(m_2 p/m_1 R_{t+1}) + \mathsf{E} \log(m_2 + m_1 R_{t+1}) - \mathsf{E} \log(m_2 + m_1 R_t).$$

In the steady state the last two terms cancel, so we get a single integral

$$\begin{aligned} a &= \log(m_2 p/m_1) - \mathsf{E} \log R_t, \\ &= \log(m_2 p/m_1) - \int dx C(x) \log x. \end{aligned} \tag{8.2.16}$$

The form (8.2.16) is very convenient for numerical computation. For analytical work it is easier to use (8.2.12) again and observe that

$$-\mathsf{E} \log R_t = -\mathsf{E} \log R_{t+1} = \log z + \mathsf{E} \log F_{t+1} + \mathsf{E} \log(1+R_t). \tag{8.2.17}$$

From (8.2.3) the reader may show that

$$\mathsf{E} \log F_t = \log n - \psi(n) \tag{8.2.18}$$

where ψ is the logarithmic derivative of the gamma function (Abramowitz and Stegum 1972, Sec. 6.3.1). From the definition of $z (= m_1^2/m_2 p)$ and (8.2.16)–(8.2.18), it follows that

$$a = \frac{1}{2} \log\left(\frac{m_2 p}{n}\right) + \mathsf{E} \log F_t + \frac{1}{2} \log(zn) + \mathsf{E} \log(1+R_t). \tag{8.2.19}$$

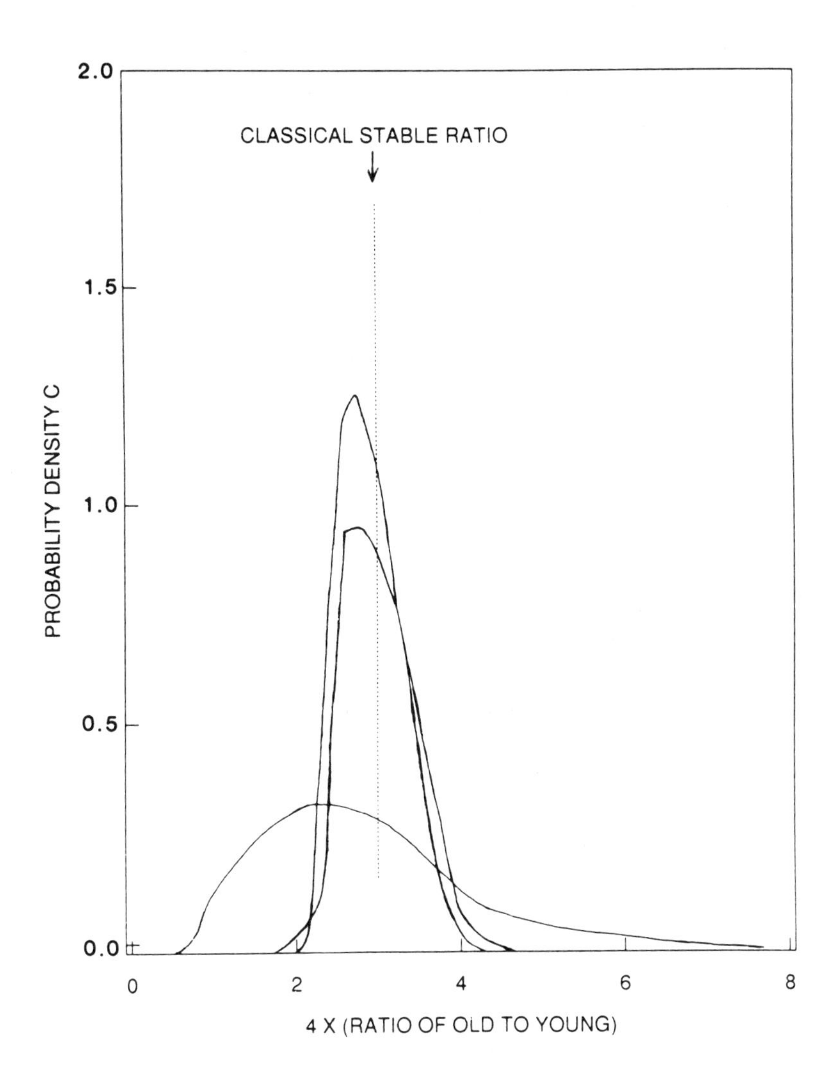

FIGURE 8.2.1. Stationary probability density of scaled age structure as defined in the text. Parameter values are $m_1 = 0.25$, $m_2 = 1$, $p = 0.75$. The classical stable ratio is indicated for vital rates fixed at these parameter values. The most spread-out density is for $n = 10$, the next for $n = 100$, and the most peaked for $n = 100$

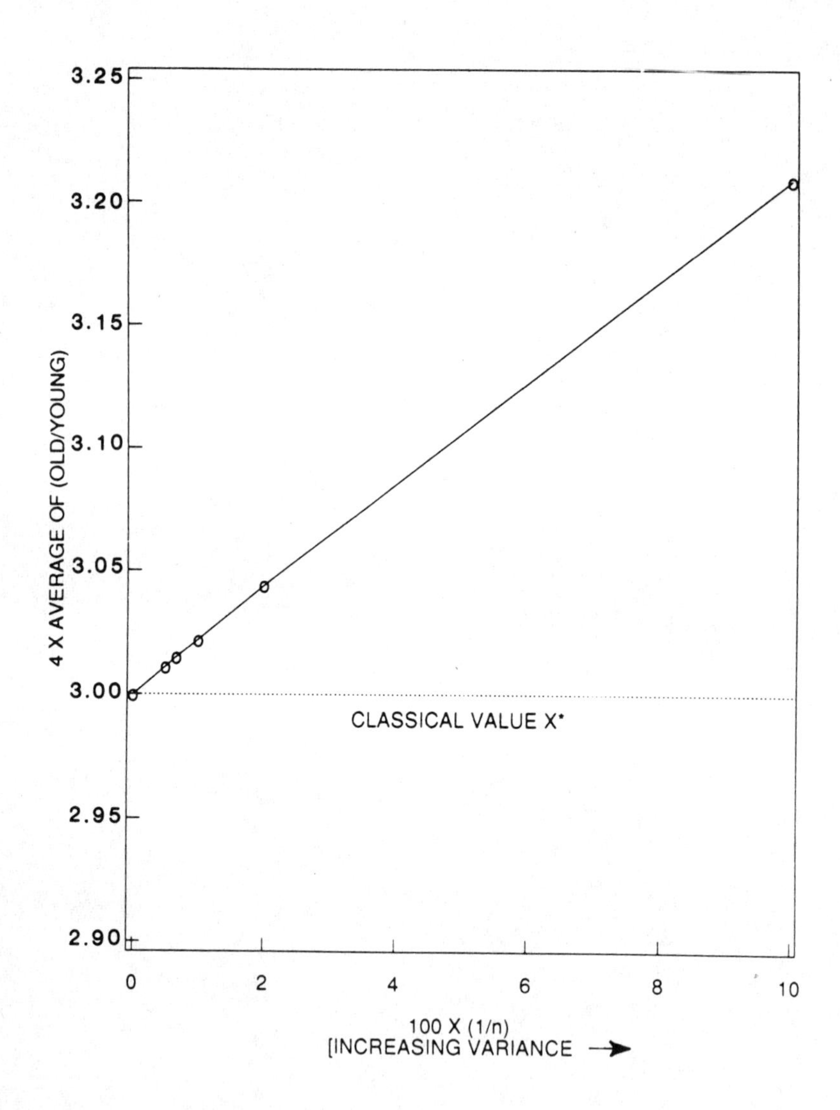

FIGURE 8.2.2. Mean age structure for increasing variance in fertility. Parameters as in preceding figure; the classical stable value is shown

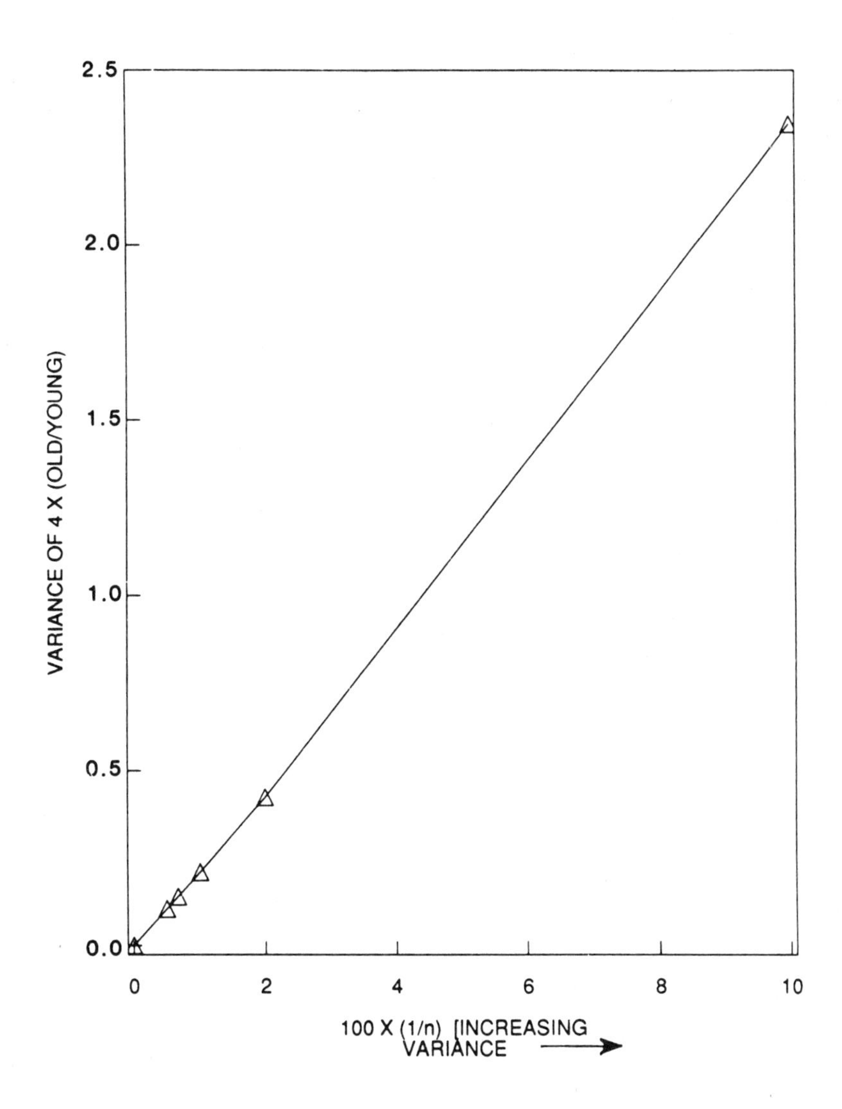

FIGURE 8.2.3. Variance in age structure for increasing variance in fertility. Parameters as in preceding figure

The last term here can be expressed (following Roerdink 1987) in terms of Kummer's confluent hypergeometric function (Abramowitz and Stegum, 1972, Sec. 13.1.3) as

$$\mathsf{E}\log(1+R_t) = \left[\frac{\partial}{\partial b}U(n,b,nz)\Big|_{b=1}\right] \Big/ U(n,1,nz), \qquad (8.2.20)$$

where

$$\Gamma(r)U(n,b,nz) = \int_0^\infty dx\, x^{n-1}(1+x)^{b-n-1}e^{-nzx}dx. \qquad (8.2.21)$$

Again, following Roerdink (1987), we obtain the final exact analytical formula

$$a = \frac{1}{2}\log(m_2 p) + \frac{1}{2}\log n - \psi(n)$$
$$+\frac{1}{2}\left\{\sum_{m=0}^{\infty}\frac{\Gamma(n+m)z_1^m}{\Gamma(1+m)m!}T_m\right\} \Big/ \left\{\sum_{m=0}^{\infty}\frac{\Gamma(n+m)z_1^m}{\Gamma(1+m)m!}N_m\right\} \qquad (8.2.22)$$

with $z_1 = nz$, and

$$T_m = \{2\psi(n) - \psi(n+m)\}\log z_1 + \{2\psi(n)[\psi(n+m) - \psi(1+m)]$$
$$2\psi(1+m)[\psi(n+m) - \psi(n)] - \Delta(n+m)\}, \qquad (8.2.23)$$

$$\Delta(b) = \left\{\frac{d^2}{db^2}\Gamma(b)\right\} \Big/ \Gamma(b),$$

$$N_m = \log z_1 + \psi(n+m) - 2\psi(1+m). \qquad (8.2.24)$$

2.4 The semelparous limit

It is interesting to consider in (8.2.1) what happens when $m_1 \to 0$ with $m_2 > 0$, $p > 0$. The average matrix (8.2.6) approaches a cyclical form,

$$\boldsymbol{b} \longrightarrow \begin{pmatrix} 0 & m_2\langle F\rangle \\ p & 0 \end{pmatrix}. \qquad (8.2.25)$$

As is well known from classical demography there is no stable age structure for this limiting $\boldsymbol{b}$, and a deterministic population with such rates would have a cyclically varying structure over time. The stochastic projection matrix is also cyclical when $m_1 = 0$,

$$\boldsymbol{X}_t \longrightarrow \begin{pmatrix} 0 & m_2 F_t \\ p & 0 \end{pmatrix}. \qquad (8.2.26)$$

Demographic ergodicity does not hold for (8.2.26); the ratio of old to young follows the equation (*cf.* (8.2.10)),

$$U_t = \frac{F_{t-1}}{F_t}U_{t-2} = \frac{F_{t-1}F_{t-3}\ldots F_{t-2m+1}}{F_tF_{t-2}\ldots F_{t-2m+2}}U_{t-2m}. \qquad (8.2.27)$$

Since the ratio above has the same number of Fs in the numerator and denominator, we have

$$\mathsf{E}\log U_t = \mathsf{E}\log U_0 = \log\left[n_0(2)/n_0(1)\right]. \tag{8.2.28}$$

The population never forgets its past when $m_1 = 0$. The stochastic growth rate when $M_1 = 0$ is easy to compute (try it) as

$$\begin{aligned} a &= \frac{1}{2}\log m_2 p + \frac{1}{2}\mathsf{E}\log F_t \\ &= \frac{1}{2}\log m_2 p + \frac{1}{2}\log n - \frac{1}{2}\psi(n). \end{aligned} \tag{8.2.29}$$

It is instructive to ask how the general formula for $m_1 > 0$, equation (8.2.22), behaves in the limit as $m_1 \to 0$. It is possible (Roerdink 1987) to expand terms in (8.2.23) to find that as $m_1 \to 0$, and thus as $z \to 0$,

$$\begin{aligned} a &\simeq \frac{1}{2}\log(m_2 p) + \frac{1}{2}\log n - \psi(n) + \frac{1}{2}\left[\psi(n) - \{\psi'(n)\log(nz)\}\right] \\ &\quad +0\left(\log(nz)^{-2}\right), \\ &= \frac{1}{2}\log(m_2 p) + \frac{1}{2}\log n - \frac{1}{2}\psi(n) - \left[\psi'(n)/\log(nz)\right], \end{aligned} \tag{8.2.30}$$

where $\psi'(x) = (d\psi/dx)$. Thus a changes very steeply with z as $z \to 0$ and in fact

$$\frac{da}{dz} \simeq \frac{\psi'(n)}{nz(\log nz)^2} \longrightarrow \infty \quad \text{as} \quad z \to 0.$$

We see that the **semelparous limit** is a **singular limit** near which a changes extremely rapidly.

2.5 CONVERGENCE RATE

Recall the Liapunov exponents of Sections 4.2.1 and 4.2.4. In the present 2-age class model there are 2 exponents, one is a and the other is ρ_2. From the general result (4.2.17) we have here

$$\begin{aligned} a + \rho_2 &= \mathsf{E}\log m_2 p F_t, \\ &= \log m_2 p + \mathsf{E}\log F_t, \\ &= \log m_2 p + \log n - \psi(n), \end{aligned} \tag{8.2.31}$$

where I use (8.2.18). Combining this result with (8.2.19) yields the exact result

$$\rho_2 = \frac{1}{2}\log\left(\frac{m_2 p}{n}\right) - \frac{1}{2}\log(zn) + \mathsf{E}\log(1 + R_t). \tag{8.2.32}$$

Recalling (8.2.24), it is clear that in the semelparous limit as $z \to 0$ (*i.e.*, $m_1 \to 0$) we have $\rho_2 \to a$, as shown more generally in Section 9.3.

9

AGE STRUCTURE: BOUNDS, GROWTH, CONVERGENCE

The analysis of age structure remains of central interest to many demographers and I present here some useful results: bounds on age structure in the presence of random rates, a relatively simple formula for growth rate a, and an analysis of periodic matrices.

1 Limits on Population Structure

An interesting question about random rates is: how much variation can they produce in population structure? By structure we mean the vector of proportions $\boldsymbol{Y}$. One way of formalizing the question is: suppose the vital rates are bounded, so that for each i, j one has either

$$(\boldsymbol{X}_t)_{ij} = 0 \quad \text{for all } t, \tag{9.1.1}$$

or

$$0 < (\boldsymbol{L})_{ij} \le (\boldsymbol{X}_t)_{ij} \le (\boldsymbol{U})_{ij} < \infty \quad \text{for all } t. \tag{9.1.2}$$

Then are there bounds such that in the statistical steady state for each t, i

$$b(i) \le Y_t(i) \le c(i)? \tag{9.1.3}$$

General results on this question are given by Seneta (1984) who provides references to earlier work. His main result (which applies more generally to products of nonnegative matrices) is that under the ergodicity conditions used here, there is a nonempty limiting set of population structures which is approached geometrically fast. However this result does not describe the limiting set.

Explicit bounds on steady state structure can be obtained for age structured populations whose Leslie matrices obey (9.1.1) except in the first row and subdiagonal. Specifically we can find an explicit equation for the vectors $\boldsymbol{c}$ and $\boldsymbol{b}$ in (9.1.3), as follows.

Label the age classes 1 through k, where k = {last age class which is reproductive}, and describe population structure by the variables $Z(i)$ defined as

$$Z_t(i) = N_t(i)/N_t(1),$$

$$\boldsymbol{Z}_t = (Z_t(2), \ldots, Z_t(k)).$$

The basic equation (4.1.2) now becomes

$$\begin{aligned} Z_{t+1}(i+1) &= S_{t+1}(i) Z_t(i) / (\boldsymbol{M}_{t+1}, \boldsymbol{W}_t) \quad i > 1, \\ &= S_{t+1}(i) Z_t(i) h(\boldsymbol{M}_{t+1}, \boldsymbol{Z}_t), \end{aligned} \tag{9.1.4}$$

where $S_t(i) = (\boldsymbol{X}_t)_{i+1,i}$, $M_t(i) = (\boldsymbol{X}_t)_{1i}$, the vectors are $\boldsymbol{S} = (S(i))$, $\boldsymbol{M} = (M(i))$, $\boldsymbol{W} = (1, \boldsymbol{Z})$, and the function $h(\boldsymbol{m}, \boldsymbol{z}) = 1/(\boldsymbol{m}, \boldsymbol{z})$. Suppose that the Ss and Ms are bounded for all t,

$$\boldsymbol{0} < \boldsymbol{s}_1 \leq \boldsymbol{S}_t \leq \boldsymbol{s}_2 < \boldsymbol{e}, \quad \boldsymbol{0} \leq \boldsymbol{m}_1 \leq \boldsymbol{M}_t \leq \boldsymbol{m}_2 < \infty. \tag{9.1.5}$$

Here and elsewhere inequalities between vectors are shorthand for inequalities which hold componentwise; similarly for matrices.

To find bounds on age structure resulting from the bounds on vital rates, observe that h in (9.1.4) is a decreasing function of $\boldsymbol{Z}$, and also that

$$h_1(\boldsymbol{Z}) = h(\boldsymbol{m}_1, \boldsymbol{Z}) \geq h(\boldsymbol{M}_t, \boldsymbol{Z}) \geq h(\boldsymbol{m}_2, \boldsymbol{Z}) = h_2(\boldsymbol{Z}). \tag{9.1.6}$$

Starting with any initial nonzero $\boldsymbol{Z}_0$ in (9.1.4) such that $\boldsymbol{0} \leq \boldsymbol{Z}_0 < \infty$, generate upper bounds as follows

$$\begin{aligned} Z_1(2) &\leq s_2(1) h_1(0) = c_1(2), \\ Z_2(3) &\leq s_2(2) c_1(2) = c_1(3), \end{aligned}$$

and so on to get a complete vector of upper bounds, $\boldsymbol{c}_1 \geq \boldsymbol{Z}_k$. These upper bounds can be used to generate a set of lower bounds,

$$\begin{aligned} Z_1(2) &\geq s_2(1) h_2(c_1) = b_1(2), \\ Z_2(3) &\geq s_2(2) h_2(c_1) = b_1(3), \end{aligned}$$

and so on to get a vector $\boldsymbol{b}_1 \leq \boldsymbol{Z}_k$. We can now iterate this procedure to generate a sequence of upper bounds $\boldsymbol{c}_m$ and lower bounds $\boldsymbol{b}_m$, $m = 1, 2, \ldots$, using the mappings

$$\begin{aligned} c_{m+1}(i+1) &= s_2(i) c_{m+1}(i) h_1(\boldsymbol{b}_m), \\ b_{m+1}(i+1) &= s_1(i) b_{m+1}(i) h_2(\boldsymbol{c}_m), \end{aligned} \tag{9.1.7}$$

where we define $b(1) = c_m(1)$ for all m, and $i = 1, \ldots, k-1$. It is easy to see that $\boldsymbol{c}_{m+1} \leq \boldsymbol{c}_m$, and $\boldsymbol{b}_{m+1} \geq \boldsymbol{b}_m$. Thus the bounds converge to a limit which can be computed by solving for an equilibrium in (9.1.7). Specifically we look for $\boldsymbol{c}$ and $\boldsymbol{b}$ which reproduce themselves under the mappings given above. A little algebra shows that if we set

$$\tau = h_1(\boldsymbol{b}),$$

$$\ell_j(i) = \begin{cases} 1 & \text{if } i = 1, \\ s_j(1)\, s_j(2) \ldots s_j(i-1) & \text{if } i > 1, \end{cases} \quad \text{for } j = 1, 2,$$

$$Q(\tau) = \sum_{x=1}^{k} \ell_2(x) m_2(x) \tau^x,$$

then we can find h as the root of the polynomial equation

$$\sum_{x=1}^{k} \ell_1(x) m_1(x) \tau^x Q^{-x-1} = 1. \tag{9.1.8}$$

In simple cases this equation yields a nice interpretation in terms of iterated products of matrices (*cf.* Section 8.1.1).

2 A Simplification for Age-Structure: Estimating a

Given Markovian vital rates, the growth rate a is in general computed according to (4.2.10) which requires an average over the joint distribution of vital rates and population structure. However, for **age-structured** populations, a remarkable simplification can be obtained as follows.

Use the scaled variables

$$Z_t(i) = N_t(i)/N_t(1)$$

of Section 9.1, and recall equations (9.1.4)–(9.1.5) and the corresponding notations. The one period growth rate is obtained by noting that

$$(\boldsymbol{e}, \boldsymbol{N}_{t+1}) = (\boldsymbol{M}_{t+1}, \boldsymbol{N}_t) + \sum_{i \geq 1} S_{t+1}(i) N_t(i) \tag{9.2.1}$$

so that

$$\begin{aligned}(\boldsymbol{e}, \boldsymbol{N}_{t+1})/(\boldsymbol{e}, \boldsymbol{N}_t) &= \left\{(\boldsymbol{M}_{t+1}, \boldsymbol{W}_t) + \textstyle\sum_{i \geq 1} S_{t+1}(i) Z_t(i)\right\} /(\boldsymbol{e}, \boldsymbol{W}_t) \\ &= (\boldsymbol{M}_{t+1}, \boldsymbol{W}_t)\left\{(\boldsymbol{e}, \boldsymbol{W}_{t+1})/(\boldsymbol{e}, \boldsymbol{W}_t)\right\}.\end{aligned}$$

Now from (9.1.4) for $i = 2$ one has $(\boldsymbol{M}_{t+1}, \boldsymbol{W}_t) = S_{t+1}(1)/Z_{t+1}(2)$ and so the logarithmic growth rate becomes

$$\begin{aligned} a &= \mathsf{E} \log\left\{(\boldsymbol{e}, \boldsymbol{N}_{t+1})/(\boldsymbol{e}, \boldsymbol{N}_t)\right\} \\ &= \mathsf{E} \log S_{t+1}(1) - \mathsf{E} \log Z_{t+1}(2) \\ &\quad + \mathsf{E} \log (\boldsymbol{e}, \boldsymbol{W}_{t+1}) - \mathsf{E} \log (\boldsymbol{e}, \boldsymbol{W}_t). \end{aligned}$$

Stationarity implies that the last two terms will cancel and so finally we get

$$a = \mathsf{E} \log S_t(1) - \mathsf{E} \log Z_t(2), \tag{9.2.2}$$

where $S_t(1)$ is the (random) survival rate of the youngest age class, and $Z_t(2) = [N_t(2)/N_t(1)]$ is the ratio of number in age-class 2 to the number in age-class 1. This equation is remarkably close to the classical analog which can be written

$$r = \log p(1) - \log[u(2)/u(1)]$$

where $u(1)$, $u(2)$ are stable proportions in age classes 1, 2. In practical cases where one has partial census information together with some statistics on $S_t(1)$, (9.2.2) yields a quick estimate of a. Equation (8.2.16) is, of course, a special case of (9.2.2) for 2 age classes.

3 Leslie Matrices with Restricted Reproduction

Ecologists and demographers (Cole 1954, Bernardelli 1941) have been interested in populations where only one or a few age classes reproduce. I consider two such cases in which demographic ergodicity does not hold.

3.1 Semelparity: One Age-Class Fertile

The Leslie matrices for such a population have the form

$$\boldsymbol{X}_t = \begin{pmatrix} 0 & \cdots & F(t) \\ P_1(t) & \cdots & 0 \\ \cdots & P_{k-1}(t) & 0 \end{pmatrix}, \tag{9.3.1}$$

where I ignore post-reproductive age classes. Let $B_t = N_t(1) =$ (births at time t). Then

$$B_t = P_{k-1}(t)P_{k-2}(t-1)\cdots P_t(t-k+1)F(t)B_{t-k}. \tag{9.3.2}$$

Iterating this and using stationarity (the right side below is independent of t) yields

$$a = (1/k)\{\mathsf{E}\log F(t) + \mathsf{E}\log P_1(t) + \cdots + \mathsf{E}\log P_{k-1}(t)\}. \tag{9.3.3}$$

(Cohen *et al.* (1983) solved the special case of a 2×2, fixed $F(t) = f$, version of (9.3.1)). For the special matrices (9.3.1) all the Liapunov exponents can be calculated as follows. Observe that the $\boldsymbol{X}_t$ in (9.3.1) can be rewritten as the product

$$\boldsymbol{X}_t = \boldsymbol{H}\boldsymbol{D}_t = \boldsymbol{H}\begin{pmatrix} P_1(t) & & & \\ & P_2(t) & & \\ & & \ddots & \\ & & & F(t) \end{pmatrix}, \tag{9.3.4}$$

$$\boldsymbol{H} = \begin{pmatrix} 0 & 0 & \cdots & \cdot & 1 \\ 1 & \cdot & \cdots & \cdot & \cdot \\ \cdot & \cdot & \cdots & \cdot & \cdot \\ 0 & \cdot & \cdots & 1 & 0 \end{pmatrix}. \tag{9.3.5}$$

The product of k of the $\boldsymbol{X}$s in (9.3.1) can also be written in this form with different entries in the diagonal matrix; thus

$$\boldsymbol{X}_k\boldsymbol{X}_{k-1}\cdots\boldsymbol{X}_1 = \boldsymbol{H}\boldsymbol{D}_{1k}, \quad \text{say.}$$

Iterating this yields

$$X_{nk}X_{nk-1}\cdots X_1 = HD_{nk} = H\begin{pmatrix} d_{nk}(1) & & \\ & \ddots & \\ & & d_{nk}(k)\end{pmatrix}$$

and since H is unitary, one has (Raghunathan 1979)

$$\rho_i = \lim_{n\to\infty}\frac{1}{nk}\log d_{nk}(i).$$

From the form of D_t one concludes that

$$\rho_2 = \cdots = \rho_k = a, \tag{9.3.6}$$

which is precisely analogous to the eigenvalue degeneracy encountered for a fixed matrix of the form (9.3.1). There is no convergence to a steady state.

Observe the remarkable fact that a is here independent of serial autocorrelation of the random process generating the vital rates. This simplicity is lost as soon as the matrix acquires another fertility.

3.2 IMPRIMITIVITY: TWO AGE-CLASSES FERTILE

Letting n, m be any two integers with $1 < n < nm = k$, consider Leslie matrices whose first row has the form

$$\begin{array}{ccc} \big(0\cdots 0 & F_1(t) & 0\cdots F_2(t)\big), \\ & \uparrow & \uparrow \\ & n\text{th} & (nm) = k\text{th} \\ & \text{column} & \text{column} \end{array} \tag{9.3.7}$$

the rest of the matrix being as in (9.3.1). The resulting Leslie matrices are irreducible but not primitive, so here too there is no convergence to a steady state for the full process. Letting B_t = (births at time t) we have

$$B_t = F_1(t)L_n(t)B_{t-n} + F_2(t)L_k(t)B_{t-k}, \tag{9.3.8}$$

with survivorships defined by

$$L_j(t) = \begin{cases} 1, & j = 1 \\ P_{j-1}(t-1)P_{j-2}(t-2)\cdots P_1(t-j+1), & 1 < j \leq k. \end{cases} \tag{9.3.9}$$

Now consider births at the times $t = n, 2n, 3n, \cdots$, and in general at $t = Tn$, T an integer. Then define a new process $Z_k = B_{kn}$ and from (9.3.7–8) get

$$Z_T = F_1(nT)L_n(nT)Z_{T-1} + F_2(nT)L_{nm}(nT)Z_{T-m}. \tag{9.3.10}$$

This new Z_T has a stable limiting behavior as $T \to \infty$ since (9.3.9) can be thought of as the birth sequence in a population described by primitive 2×2 Leslie matrices. Hence the degeneracy of (9.3.5) disappears under a suitable time scaling, and one has the strict inequality $a > \rho_2$ since weak ergodicity holds for the scaled 2×2 matrices.

10

SYNERGY, CONSTRAINTS, CONVEXITY

This book deals with many of the differences between the multidimensional dynamics of structured populations and those of scalar growth models. It is important to identify those features of the structured case which differ markedly from the scalar case. The first two sections below show that the differences can be quite considerable. Section 1 discusses the effect of autocorrelation. Section 2 presents an example where serially uncorrelated random variation raises the growth rate of a population above its possible deterministic growth rate. The third section summarizes a potentially useful result of Cohen concerning the parametric sensitivity of stochastic rates. The last section shows how strong constraints on population vital rates can lead to "scalar" behavior.

1 Autocorrelation

Consider the general projection model $\boldsymbol{N}_{t+1} = \boldsymbol{X}_{t+1}\boldsymbol{N}_t$ and suppose that $\boldsymbol{b} = \mathsf{E}\,\boldsymbol{X}_t$. There are 3 growth rates relevant to this model. One is a as defined in Section 4.2.1. The second is the growth rate of the average population (*cf.* Chapter 7),

$$\log \mu = \lim_{t\to\infty} \frac{1}{t} \log \mathsf{E}\, M_t. \tag{10.1.1}$$

The third is the deterministic growth rate given by using the average vital rates, $r_0 = \log \lambda_0 = \log$ (dominant eigenvalue of $\boldsymbol{b}$). We know (Jensen's inequality) that

$$a \le \log \mu; \tag{10.1.2}$$

thus, when vital rates are I.I.D. (Section 4.1) we have that

$$a \le \log \mu = r_0. \tag{10.1.3}$$

In general (*i.e.*, with Markov rates, Chapter 7) serial autocorrelation can complicate matters and we may have

$$a < r_0 < \log \mu \tag{10.1.4}$$

or

$$r_0 < a < \log \mu. \tag{10.1.5}$$

Cohen (1979) provides numerical examples. These effects of autocorrelation are unique to multidimensional structured population models.

2 Synergistic Effects of Environment

Even when there is *no* autocorrelation, structured populations can behave dramatically differently from scalar ones. A striking illustration is provided by situations in which the randomness of the environment makes the difference between population growth and decline.

Consider a population with 2 age classes and suppose that the population's vital rates are given by one of 2 Leslie matrices,

$$\boldsymbol{A} = \begin{pmatrix} 1/4 & 3-x \\ 1/4 & 0 \end{pmatrix}, \tag{10.2.1}$$

or

$$\boldsymbol{B} = \begin{pmatrix} 15/16 - 0.1 & 1/16 + 0.09 \\ 1 & 0 \end{pmatrix}. \tag{10.2.2}$$

Take the environment to change randomly so that in each time interval the population's Leslie matrix is $\boldsymbol{A}$ with probability of p or $\boldsymbol{B}$ with probability $(1-p)$. If we set $x = 0.142857$ and $p = 0.5$ a numerical simulation (of 5000 iterations of the stochastic growth process) yields an estimate $a = +0.1954$ with a sample standard error of $\widehat{s} = 0.0047$. Thus, the population should increase with probability one over the long term. Yet computation of dominant eigenvalues of the matrices shows

$$\begin{aligned} \log \lambda_0(\boldsymbol{A}) &= -0.0209, \\ \log \lambda_0(\boldsymbol{B}) &= -0.0087. \end{aligned} \tag{10.2.3}$$

Therefore in the absence of a random environment with matrix $\boldsymbol{A}$ or $\boldsymbol{B}$ fixed forever, the population would decline. We have here a synergistic effect of random variation. Key (1986) suggested this term in the context of a multitype branching process using special kinds of matrices. The example given above is easily generalized to more parameters or dimensions.

An equally important aspect of this example is that it shows decisively that the logarithmic mean dominant eigenvalue (of the underlying matrices) *cannot* accurately describe a, since

$$LM = 0.5[\log \lambda_0(\boldsymbol{A}) + \log \lambda_0(\boldsymbol{B})] < 0, \tag{10.2.4}$$

whereas $a > 0$. It should also be obvious that this example derives fundamentally from the multidimensional character of the problem, and would not be possible without age structure.

The reader who is curious about the provenance of (10.2.1) and (10.2.2) should note that I started with simple rational members as entries. The phenomenon above is not special in any numerical sense but occurs over a range of values of x. Finally it is possible to deduce a general rule for constructing such examples with many age classes (although I will not discuss it here).

3 Convexity Properties

The sensitivity of growth rate to parameters affecting vital rates is important in classical demography. Results in Chapter 11 will provide some information on the sensitivity of a to stochastic properties of vital rates. There are many problems in which the vital rates depend on parameters and one wants to know how changes in these parameters will affect a. Cohen (1980) has established the following general result which provides some information.

Let $(\theta_1, \theta_2, \ldots, \theta_m) = \boldsymbol{\theta}$ be a set of real parameters, and suppose that the random vital rates depend on these:

$$(\boldsymbol{X}_t)_{ij} = F_{tij}(\boldsymbol{\theta}). \tag{10.3.1}$$

Here the F_{tij} are random functions, some of which are identically zero. *Assume* that for every t, i, j:

$$\begin{array}{ll} \textit{either} & F_{tij}(\boldsymbol{\theta}) = 0, \\ \textit{or} & \log F_{tij}(\boldsymbol{\theta}) \text{ is a convex function of } \boldsymbol{\theta}. \end{array} \tag{10.3.2}$$

Then a defined via (4.2.2) for each $\boldsymbol{\theta}$ is a function $a(\boldsymbol{\theta})$, and Cohen proves that $a(\boldsymbol{\theta})$ is a convex function of $\boldsymbol{\theta}$. A similar result holds for the $\log \mu$ defined in (10.1.1).

4 Leslie Matrices with Constraints

Kim and Sykes (1978) studied an interesting class of Leslie matrices with time-dependent vital rates. They argue that density or other autoregulation will constrain vital rates so that the net reproductive rate of the population is always unity. Both the form of their constraint and their argument can be greatly generalized as follows. Consider a collection of vital rate matrices $\boldsymbol{A}_1, \boldsymbol{A}_2 \ldots$; the collection *shares reproductive value* if there is a positive vector $\boldsymbol{v}$ and positive numbers $\lambda_1, \lambda_2, \ldots$ such that

$$\boldsymbol{v}^T \boldsymbol{A}_i = \lambda_i \boldsymbol{v}^T, \quad i = 1, 2, \ldots, \quad |v| < \infty, \tag{10.4.1}$$

where superscript T means transpose. Alternatively the collection *shares stable structure* if there is a positive vector $\boldsymbol{u}$ and positive numbers $\lambda_1, \lambda_2, \ldots$ such that

$$\boldsymbol{A}_i \boldsymbol{u} = \lambda_i \boldsymbol{u}, \quad i = 1, 2, \ldots, \quad |\boldsymbol{u}| < \infty. \tag{10.4.2}$$

Suppose now that there is a stationary stochastic process which chooses the $\boldsymbol{X}$'s in (4.2.1) from a collection of matrices of type (10.4.1) or (10.4.2). Take specifically the type (10.4.1) and observe that an initial vector $\boldsymbol{N}_0$ becomes

$$\boldsymbol{N}_t = \boldsymbol{A}_{i_t} \boldsymbol{A}_{i_{t-1}} \cdots \boldsymbol{A}_{i_1} \boldsymbol{N}_0, \tag{10.4.3}$$

where $i_0, i_1, \ldots$ are integers which depend on the underlying stochastic process. Now using (10.4.1) yields

$$v^T N_t = \lambda_{i_t} \lambda_{i_{T-1}} \cdots \lambda_{i_1} v^T N_0, \tag{10.4.4}$$

and so

$$\begin{aligned} a &= \lim_{t\to\infty} \frac{1}{t} \log (e, N_t) \\ &= \lim_{t\to\infty} \frac{1}{t} \log (v^T N_t) \quad \text{(because } |v| \text{ is bounded)} \\ &= \lim_{t\to\infty} \frac{1}{t} \sum_{m=1}^{t} \log \lambda_{i_m} \\ &= \mathsf{E} \log \lambda, \end{aligned} \tag{10.4.5}$$

where the expectation is over the underlying stochastic process.

There are some remarkable aspects of (10.4.5). Under the stationary rate assumptions, the average $\mathsf{E}(\log \lambda)$ is *independent* of serial autocorrelation in the vital rate sequence. With respect to growth rate, the product in (10.4.3) behaves as if the matrices commuted, which they do not except in very special cases. Therefore the conditions (10.4.1) or (10.4.2) result in an effectively one-dimensional growth rate sequence for an otherwise complex age-structured process (the relationship between one-period and long-run growth rates in (10.4.5) is generally true only for scalar processes). For example, the age-structure in (10.4.3) under condition (10.4.1) does *not* converge to a degenerate (single vector) limit. It is easy to see this from Kim and Syke's work on 2 age-classes with time-dependent rates. Similarly in the case of (10.4.2), although age-structure does converge to the degenerate limit u, the reproductive values do not.

These effects are brought about by the internal correlation structure of the vital rates imposed by (10.4.1) or (10.4.2). It is instructive to construct some examples. First take case (10.4.1) for 3 age-classes and let

$$v^T = (1, 1+t, 1-s), \quad t > 0, \quad s > 0, \tag{10.4.6}$$

so that reproductive value is peaked at age class 2. Set

$$A_i = \begin{pmatrix} m_{1i} & m_{2i} & m_{3i} \\ p_{1i} & 0 & 0 \\ 0 & p_{2i} & 0 \end{pmatrix}, \tag{10.4.7}$$

choose positive numbers λ_i, and observe that (10.4.1) implies

$$\begin{aligned} m_{1i} &= \lambda_i - (1+t)p_{1i}, \\ m_{2i} &= (1+t)\lambda_i - (1-s)p_{2i}, \\ m_{3i} &= \lambda_i(1-s). \end{aligned} \tag{10.4.8}$$

For example, pick $t = s = 0.2$ in (10.4.6), $\lambda_1 = 1.1$, $\lambda_2 = 1.2$, $\lambda_3 = 1.3$, and $p_{i1} = 0.8$, $p_{i2} = 0.6$ for $i = 1, 2, 3$ to get

$$A_1 = \begin{pmatrix} 0.14 & 0.84 & 0.88 \\ 0.8 & 0 & 0 \\ 0 & 0.6 & 0 \end{pmatrix},$$

$$A_2 = \begin{pmatrix} 0.24 & 0.96 & 0.96 \\ 0.8 & 0 & 0 \\ 0 & 0.6 & 0 \end{pmatrix}, \tag{10.4.9}$$

$$A_3 = \begin{pmatrix} 0.34 & 1.08 & 1.04 \\ 0.8 & 0 & 0 \\ 0 & 0.6 & 0 \end{pmatrix}.$$

It is easily verified that $A_1 A_2 \neq A_2 A_1$ and in general that the above matrices do not commute. A little more algebra would produce a set which did not share survival rates. Yet the growth rate of a population whose vital rates were chosen from the set $\{A_1, A_2, A_3\}$ according to a stationary process (as in Section 4.1) would have a given by (10.4.5). In the present case if $\pi_i =$ (long-run frequency of choosing A_1),

$$a = \sum_{i=1}^{3} \pi_i \log \lambda_i. \tag{10.4.10}$$

A similar construction is easily made starting from (10.4.2).

The general pattern of (10.4.1) is that at every age, fertility and survival rate are inversely correlated in each time interval. On the other hand, (10.4.2) fixes survival rates uniquely and results in negative correlations between fertility at all ages in each time interval. In terms of possible regulatory mechanisms these differences are substantial.

As a final and important note, these situations in which growth rate (and only growth rate!) behaves in a "scalar-model" way, are very complex biologically. They involve very precise relationships between and within age classes in their response to changing environments.

11

SENSITIVITY ANALYSIS OF GROWTH RATE

A central question in demography is: how do vital rates interact to determine growth rate and population structure? The answer is a bit complicated even in the deterministic case, because there is no analytical formula giving r in terms of vital rates. Ecologists and demographers have resorted to studying how r changes when vital rates change (Lewontin 1965, Keyfitz 1968, 1977, Caswell 1978, Arthur 1980). This amounts to studying the derivatives of r, and is called sensitivity analysis (Caswell 1978). The stochastic analog is the sensitivity analysis of the long-run growth rate a. In this chapter I begin with a quick summary of sensitivity analysis for r, and then set out the sensitivity analysis for a.

1 Deterministic Sensitivity Analysis

In the classical model with a fixed matrix $\boldsymbol{b}$ of vital rates, the long-run growth rate $r_0 = \log \lambda_0$ where λ_0 is the dominant eigenvalue of $\boldsymbol{b}$ (as in Section 2.1). We want to know how r_0 changes when the elements of $\boldsymbol{b}$ are changed. This question is the subject of the perturbation theory of matrices (Kato 1966) but I think it is useful to sketch the analysis here. I assume that $\boldsymbol{b}$ is primitive and has a full spectral decomposition (as in Section 2.2).

Suppose we change vital rates so that matrix $\boldsymbol{b}$ is replaced by $\boldsymbol{b}+\epsilon\boldsymbol{c}$, with ϵ being a small ($|\epsilon| \ll 1$) parameter. The new matrix will have dominant eigenvalue $\lambda_0(\epsilon)$, say, and corresponding right eigenvector $\boldsymbol{u}_0(\epsilon)$,

$$(\boldsymbol{b} + \epsilon\boldsymbol{c})\boldsymbol{u}_0(\epsilon) = \lambda_0(\epsilon)\, \boldsymbol{u}_0(\epsilon). \tag{11.1.1}$$

When $\epsilon = 0$, $\boldsymbol{u}_0(0) = \boldsymbol{u}_0$ and $\lambda_0(0) = \lambda_0$. When $\epsilon \neq 0$ it is possible to write

$$\begin{aligned} u_0(\epsilon) &= \boldsymbol{u}_0 + \epsilon\boldsymbol{u}_1 + \epsilon^2\boldsymbol{u}_2 + \cdots, \\ \lambda_0(\epsilon) &= \lambda_0 + \epsilon\lambda_1 + \epsilon^2\lambda_2 + \cdots. \end{aligned} \tag{11.1.2}$$

Furthermore we can require (why?) that

$$\boldsymbol{v}_0^T\boldsymbol{u}_i = 0 \quad \text{for} \quad i = 1, 2, \ldots, \tag{11.1.3}$$

where $\boldsymbol{v}_0$ is the left eigenvector of $\boldsymbol{b}$ corresponding to $\boldsymbol{u}_0$. Inserting (11.1.2) into (11.1.1) and equating the coefficients of $\epsilon^0, \epsilon^1, \epsilon^2$, *etc.*g, on the left to corresponding coefficients on the right of that equation, the ϵ-coefficients yield

$$\boldsymbol{b}\boldsymbol{u}_1 + \boldsymbol{c}\boldsymbol{u}_0 = \lambda_1 \boldsymbol{u}_0 + \lambda_0 \boldsymbol{u}_1. \tag{11.1.4}$$

Taking the scalar product of $\boldsymbol{v}_0$ on both sides and using (11.1.3) we get

$$\lambda_1 = (\boldsymbol{v}_0^T \boldsymbol{c}\boldsymbol{u}_0)/\boldsymbol{v}_0^T \boldsymbol{u}_0). \tag{11.1.5}$$

If we were interested only in changing one vital rate, say b_{ij} to $b_{ij} + \epsilon c_{ij}$, then

$$\frac{\partial \lambda_0}{\partial b_{ij}} = \lim_{(\epsilon c_{ij} \to 0)} \frac{\epsilon \lambda_1}{\epsilon c_{ij}} = \frac{v_0(i) u_0(j)}{(\boldsymbol{v}_0^T \boldsymbol{u}_0)}. \tag{11.1.6}$$

This is Caswell's (1978) formula (and also the standard result of perturbation theory) and has direct ecological applications. It is easy to show in the same way that for any eigenvalue λ_α of $\boldsymbol{b}$ we have

$$\frac{\partial \lambda_\alpha}{\partial b_{ij}} = \frac{v_\alpha(i) u_\alpha(j)}{(\boldsymbol{v}_\alpha^T \boldsymbol{u}_\alpha)}, \tag{11.1.7}$$

although this formula has little been used in demography (see Caswell 1989).

Finally, note that the above technique yields the change in the population structure vector. From (11.1.2) changing $\boldsymbol{b}$ causes a change of $\epsilon \boldsymbol{u}_1$ (to order ϵ) in $\boldsymbol{u}_0$. From (11.1.3) we can write the expansion

$$\boldsymbol{u}_1 = \sum_{\alpha \neq 0} c_\alpha \boldsymbol{u}_\alpha, \tag{11.1.8}$$

with the c_α being unknown coefficients, and where the index $\alpha = 1, \ldots, k-1$, for the non-dominant eigenvalues and corresponding eigenvectors of $\boldsymbol{b}$. The reader may show by analogy with (11.1.4–5) that

$$\boldsymbol{u}_1 = \sum_{\alpha \neq 0} \frac{(\boldsymbol{v}_\alpha^T \boldsymbol{c}\boldsymbol{u}_0)}{(\lambda_0 - \lambda_\alpha)(\boldsymbol{v}_\alpha^t \boldsymbol{u}_\alpha)}. \tag{11.1.9}$$

The perturbation calculations given above underlie the similar but more involved results in Sections 3.2 and 7.4. The stochastic situation requires a different approach, as given below.

2 Stochastic Sensitivity Analysis

I now present stochastic results which extend the deterministic ones above. The discussion here assumes the necessary analyticity of a, which was established by Ruelle (1979). I suppose that the rather general Assumptions 4.2.1

and 4.2.2 of Chapter 4 apply (demographic ergodicity, stationary stochastic process). In the calculation to follow the various quantities refer to the statistically stationary state.

We begin with the random sequences $\{\boldsymbol{X}_i\}, \{\boldsymbol{U}_i\}, \{\boldsymbol{V}\}$ of vital rates, population structure and reproductive value respectively such that:

$$\boldsymbol{U}_{i+1} = \frac{\boldsymbol{X}_{i+1}\boldsymbol{U}_i}{\lambda(i+1)} \quad \text{with} \quad \lambda(i) = (\boldsymbol{e}, \boldsymbol{X}_i\boldsymbol{U}_{i-1}),$$

and

$$\boldsymbol{V}_i^T = \frac{\boldsymbol{V}_{i+1}^T\boldsymbol{X}_{i+1}}{(\boldsymbol{e}, \boldsymbol{V}_{i+1}^T\boldsymbol{X}_{i+1})}. \tag{11.2.1}$$

The long-run growth rate may be computed as

$$a = \lim_{n\to\infty} \frac{1}{n} \log \alpha_n, \tag{11.2.2}$$

with

$$\alpha_n = (\boldsymbol{V}_n^T\boldsymbol{X}_n\boldsymbol{X}_{n-1}\cdots\boldsymbol{X}_1\boldsymbol{U}_0). \tag{11.2.3}$$

Suppose now that we perturb the vital rates. In the stochastic case this means that the entire vital-rate sequence $\{\boldsymbol{X}_i\}$ is changed to, say, $\{\boldsymbol{X}_i + \epsilon\boldsymbol{C}_i\} = \boldsymbol{Z}_i(\epsilon)$. The long-run growth rate of this new sequence may again be computed as

$$a(\epsilon) = \lim_{n\to\infty} \frac{1}{n} \log \alpha_n(\epsilon), \tag{11.2.4}$$

with $\alpha_n(\epsilon) = (\boldsymbol{V}_n^T\boldsymbol{Z}_n(\epsilon)\boldsymbol{Z}_{n-1}(\epsilon)\cdots\boldsymbol{Z}_1(\epsilon)\boldsymbol{U}_0)$. (We can use the same vectors because of demographic weak ergodicity.) Now

$$\alpha_n(\epsilon) = \alpha_n + \epsilon\boldsymbol{V}_n^T\left(\sum_{i=1}^n \boldsymbol{X}_n\ldots\boldsymbol{C}_i\boldsymbol{X}_{i-1}\ldots\boldsymbol{X}_1\right)\boldsymbol{U}_0 + 0(\epsilon^2). \tag{11.2.5}$$

Thus we have that

$$a(\epsilon) = a + \epsilon a_1 + 0(\epsilon^2), \tag{11.2.6}$$

and we compute a_1 as follows. From (11.2.4–6),

$$\begin{aligned}
a_1 &= \lim_{n\to\infty} \frac{1}{n} \frac{\boldsymbol{V}_n^T\left(\sum_i \boldsymbol{X}_n\ldots\boldsymbol{X}_{i+1}\boldsymbol{C}_i\boldsymbol{X}_{i-1}\ldots\boldsymbol{X}_1\right)\boldsymbol{U}_0}{(\boldsymbol{V}_n^T\boldsymbol{X}_n\boldsymbol{X}_{n-1}\ldots\boldsymbol{X}_i\ldots\boldsymbol{X}_1)\boldsymbol{U}_0} \\
&= \lim_{n\to\infty} \frac{1}{n} \sum \frac{(\boldsymbol{V}_i^T\boldsymbol{C}_i\boldsymbol{U}_{i-1})}{(\boldsymbol{V}_i^T\boldsymbol{X}_i\boldsymbol{U}_{i-1})} \\
&= \lim_{n\to\infty} \frac{1}{n} \sum \frac{(\boldsymbol{V}_i^T\boldsymbol{C}_i\boldsymbol{U}_{i-1})}{\lambda(i)(\boldsymbol{V}_i^T\boldsymbol{U}_i)} \qquad (\text{ using 11.2.1}) \\
&= \mathsf{E}\left\{\frac{\boldsymbol{V}_t^T\boldsymbol{C}_t\boldsymbol{U}_{t-1}}{\lambda(t)\boldsymbol{V}_t^T\boldsymbol{U}_t}\right\}.
\end{aligned} \tag{11.2.7}$$

This equation gives the change in a. If we are only concerned with perturbing one vital rate, say the ij element of the vital rate matrix from $X_{t,ij}$ to $(X_{t,ij} + C_{t,ij})$, then for small C the infinitesimal change in a will be

$$\Delta a = \mathsf{E} \left\{ \frac{V_t(i)U_t(j)C_{t,ij}}{\lambda(t)\boldsymbol{V}^T(t)\boldsymbol{U}(t)} \right\}. \tag{11.2.8}$$

This stochastic sensitivity formula is a generalized analog of Caswell's result. It is easy to check that (11.2.8) reduces to the deterministic result if there is no randomness.

3 Applications

The sensitivity results are essential in the study of evolutionary stable states (ESSs). An ESS is an optimal phenotype among some constrained set of possible phenotypes, and is identified by the condition that no other phenotype can successfully compete with it. In density-independent models this means that we look for the phenotype which has the maximum growth rate. As an example, consider the problem (Charnov 1982, 1988) of determining the ESS for allocation of resources to male versus female offspring in a population of simultaneous hermaphrodites. Tuljapurkar (1989) shows that for age-structured populations the stochastic ESS is found using an appropriate version of (11.2.7). Other applications of this sort are possible.

More generally, there has been little direct analysis of the numerical and qualitative properties of (11.2.7). Some direct analysis has been done in the study of life histories (Chapter 15) but the sort of information that exists in the deterministic case is needed.

12

GROWTH RATES FOR SMALL NOISE

The long-run growth rate a is central to questions of evolution (Chapter 6), prediction (Chapter 14) and extinction. However, it is only useful if we can describe how vital rates and uncertainty determine a. This is difficult because there is no general formula to compute a for arbitrary vital rates and variability. In addition, the exactly known cases of a (Chapter 8) do not generalize; worse, they reveal singular behavior near parameter limits where demographic ergodicity is lost. One useful and general approach is to develop a systematic approximation to a when the magnitude of random variation is small. This was done by Tuljapurkar (1982b) and the results have since been applied to a number of ecological and demographic problems. The method and some extensions are presented below.

1 Second-Order Expansion of a for General Matrices

In the random-rates model write the matrix of vital rates as a sum of two parts

$$\boldsymbol{X}_t = \boldsymbol{b} + \epsilon \boldsymbol{H}_t, \tag{12.1.1}$$

where

$$\boldsymbol{b} = \mathsf{E}(\boldsymbol{X}_t) \quad \text{and} \quad \mathsf{E}(\boldsymbol{H}_t) = 0. \tag{12.1.2}$$

The average matrix $\boldsymbol{b}$ is assumed to be primitive and have a simple spectral decomposition (as in Section 2.2). As usual, let $\lambda_0, \boldsymbol{u}, \boldsymbol{v}$ be the dominant eigenvalue and corresponding right and left eigenvectors of $\boldsymbol{b}$. The parameter ϵ in (12.1.1) measures the magnitude of random variation since the nth moment of $(\boldsymbol{X}_t - \boldsymbol{b})$ is proportional to ϵ^n. The temporal character of the vital rates is described by moments of variation at one time, such as

$$\boldsymbol{c}(0) = \mathsf{E}(\boldsymbol{H}_t \otimes \boldsymbol{H}_t), \tag{12.1.3}$$

and by covariances across time, such as

$$\boldsymbol{c}(\ell) = \mathsf{E}(\boldsymbol{H}_t \otimes \boldsymbol{H}_{t+\ell}), \quad \ell > 0. \tag{12.1.4}$$

I assume (see Chapter 4) demographic ergodicity and that $\boldsymbol{H}_t$ is a stationary, rapidly mixing stochastic process.

We begin by observing that the stochastic growth rate can be computed as

$$\begin{aligned} a &= \lim_{t\to\infty} \frac{1}{t} \mathsf{E} \log(\boldsymbol{e}, \boldsymbol{X}_t\boldsymbol{X}_{t-1}\ldots\boldsymbol{X}_1\boldsymbol{u}) \\ &= \lim_{t\to\infty} \frac{1}{t} \mathsf{E} \log(\boldsymbol{v}, \boldsymbol{X}_t\boldsymbol{X}_{t-1}\ldots\boldsymbol{X}_1\boldsymbol{u}). \end{aligned} \tag{12.1.5}$$

The matrix product in (12.1.5) is now expanded by using (12.1.1) to get

$$\boldsymbol{X}_t\boldsymbol{X}_{t-1}\ldots\boldsymbol{X}_1 = (\boldsymbol{b}+\epsilon\boldsymbol{H}_t)(\boldsymbol{b}+\epsilon\boldsymbol{H}_{t-1})\ldots(\boldsymbol{b}+\epsilon\boldsymbol{H}_1) \tag{12.1.6}$$

$$\begin{aligned} &= \boldsymbol{b}^t + \epsilon \sum_{i=1}^{t} \boldsymbol{b}^{t-i}\boldsymbol{H}_i\boldsymbol{b}^{i-1} \\ &\quad + \epsilon^2 \sum_{i=1}^{t-1}\sum_{j=1}^{t-i} \boldsymbol{b}^{t-i-j}\boldsymbol{H}_{i+j}\boldsymbol{b}^{j-1}\boldsymbol{H}_i\boldsymbol{b}^{i-1} + O(\epsilon^3) \\ &= \boldsymbol{b}^t + \epsilon\boldsymbol{S}_{1t} + \epsilon^2\boldsymbol{S}_{2t} + 0(\epsilon^3). \end{aligned} \tag{12.1.7}$$

Next compute

$$\begin{aligned} &\log(\boldsymbol{v}, [\boldsymbol{b}^t + \epsilon\boldsymbol{S}_{1t} + \epsilon^2\boldsymbol{S}_{2t} + O(\epsilon^3)]\boldsymbol{u}) \\ &= \log(\boldsymbol{v}, \boldsymbol{b}^t\boldsymbol{u}) + \log\left[1 + \epsilon\frac{(\boldsymbol{v}, \boldsymbol{S}_{1t}\boldsymbol{u})}{(\boldsymbol{v}, \boldsymbol{b}^t\boldsymbol{u})} + \epsilon^2\frac{(\boldsymbol{v}, \boldsymbol{S}_{2t}\boldsymbol{u})}{(\boldsymbol{v}, \boldsymbol{b}^t\boldsymbol{u})} + 0(\epsilon^3)\right] \end{aligned} \tag{12.1.8}$$

Using the fact that

$$(\boldsymbol{v}, \boldsymbol{b}^t\boldsymbol{u}) = \lambda_0^t(\boldsymbol{v}, \boldsymbol{u}), \tag{12.1.9}$$

and letting $(\boldsymbol{v}, \boldsymbol{u}) = 1$ as in Chapter 2, continue with (12.1.8) to get

$$\begin{aligned} \log(\boldsymbol{v}, \boldsymbol{X}_t\ldots\boldsymbol{X}_1\boldsymbol{u}) &= t\log\lambda_0 + \epsilon\frac{(\boldsymbol{v}, \boldsymbol{S}_{1t}\boldsymbol{u})}{\lambda_0^t} + \epsilon^2\frac{(\boldsymbol{v}, \boldsymbol{S}_{2t}\boldsymbol{u})}{\lambda_0^t} \\ &\quad - \epsilon^2\frac{(\boldsymbol{v}, \boldsymbol{S}_{1t}\boldsymbol{u})^2}{2\lambda_0^{2t}} + O(\epsilon^3). \end{aligned} \tag{12.1.10}$$

To use (12.1.10) in (12.1.5) we need the following computations

$$\frac{(\boldsymbol{v}, \boldsymbol{S}_{1t}\boldsymbol{u})}{\lambda_0^t} = \sum_{i=1}^{t} \frac{(\boldsymbol{v}, \boldsymbol{H}_i\boldsymbol{u})}{\lambda_0}, \tag{12.1.11}$$

$$\frac{(\boldsymbol{v}, \boldsymbol{S}_{1t}\boldsymbol{u})^2}{\lambda_0^{2t}} = \sum_{i=1}^{t} \frac{(\boldsymbol{v}, \boldsymbol{H}_i\boldsymbol{u})^2}{\lambda_0^2} + \sum_{\substack{i=1 \\ i\neq j}}^{t}\sum_{j=1}^{t} \frac{(\boldsymbol{v}, \boldsymbol{H}_i\boldsymbol{u})(\boldsymbol{v}, \boldsymbol{H}_j\boldsymbol{u})}{\lambda_0^2}, \tag{12.1.12}$$

$$\frac{(\boldsymbol{v}, \boldsymbol{S}_{2t}\boldsymbol{u})}{\lambda_0^t} = \sum_{i=1}^{t-1}\sum_{j=1}^{t-i} \frac{(\boldsymbol{v}, \boldsymbol{H}_{i+j}\boldsymbol{b}^{j-1}\boldsymbol{H}_i\boldsymbol{u})}{\lambda_0^{j+1}}. \tag{12.1.13}$$

In (12.1.13) insert the simple spectral decomposition (2.2.10) which says that

$$\boldsymbol{b}^m = \lambda_0^m(\boldsymbol{u}\boldsymbol{v}^T + \boldsymbol{q}^m), \quad m \geq 1, \tag{12.1.14}$$

and get

$$\begin{aligned}
\frac{(\boldsymbol{v}, \boldsymbol{S}_{2t}\boldsymbol{u})}{\lambda_0^t} &= \sum_{i=1}^{t-1} \frac{(\boldsymbol{v}, \boldsymbol{H}_{i+1}\boldsymbol{H}_i\boldsymbol{u})}{\lambda_0^2} + \sum_{i=1}^{t-1}\sum_{j=2}^{t-i} \frac{(\boldsymbol{v}, \boldsymbol{H}_{i+j}\boldsymbol{u})(\boldsymbol{v}, \boldsymbol{H}_i\boldsymbol{u})}{\lambda_0^2} \\
&\quad + \sum_{i=1}^{t-1}\sum_{j=2}^{t-i} \frac{(\boldsymbol{v}, \boldsymbol{H}_{i+j}\boldsymbol{q}^{j-1}\boldsymbol{H}_i\boldsymbol{u})}{\lambda_0^2} \\
&= \sum_{i=1}^{t-1}\sum_{j=1}^{t-i} \frac{(\boldsymbol{v}, \boldsymbol{H}_{i+j}\boldsymbol{u})(\boldsymbol{v}\boldsymbol{H}_i\boldsymbol{u})}{\lambda_0^2} + \sum_{i=1}^{t-1}\sum_{j=1}^{t-i} \frac{(\boldsymbol{v}, \boldsymbol{H}_{i+j}\boldsymbol{q}^{j-1}\boldsymbol{H}_i\boldsymbol{u})}{\lambda_0^2} \\
&\quad - \sum_{i=1}^{t-1} \frac{(\boldsymbol{v}, \boldsymbol{H}_{i+1}\boldsymbol{u})(\boldsymbol{v}, \boldsymbol{H}_i\boldsymbol{u})}{\lambda_0^2}.
\end{aligned} \tag{12.1.15}$$

Between the first and second steps we use the fact that $\boldsymbol{I} = \boldsymbol{u}\boldsymbol{v}^T + (\boldsymbol{I} - \boldsymbol{u}\boldsymbol{v}^T)$ where $\boldsymbol{I}$ is the identity matrix.

Now put (12.1.11)–(12.1.15) together into (12.1.5) and take the limit and expectation, using stationarity. The following intermediate steps are worth checking:

$$\lim_{t\to\infty} \frac{1}{t}\mathsf{E}\sum_{i=1}^{t} \frac{(\boldsymbol{v}, \boldsymbol{H}_i\boldsymbol{u})^2}{\lambda_0^2} = \frac{(\boldsymbol{v}\otimes\boldsymbol{v})^T\boldsymbol{c}(0)(\boldsymbol{u}\otimes\boldsymbol{u})}{\lambda_0^2},$$

$$\lim_{t\to\infty} \frac{1}{t}\mathsf{E}\sum_{i=1}^{t-1}\sum_{j=1}^{t-1} \frac{(\boldsymbol{v}, \boldsymbol{H}_{i+j}\boldsymbol{q}^{j-1}\boldsymbol{H}_i\boldsymbol{u})}{\lambda_0^2} = \sum_{j=1}^{\infty}\mathsf{E}\,(\boldsymbol{v}, \boldsymbol{H}_{j+1}\boldsymbol{q}^{j-1}\boldsymbol{H}_1\boldsymbol{u}).$$

The final expression for a up to second order in ϵ is

$$\begin{aligned}
a &\cong \log\lambda_0 - \epsilon^2\frac{(\boldsymbol{v}\otimes\boldsymbol{v})^T\boldsymbol{c}(0)(\boldsymbol{u}\otimes\boldsymbol{u})}{2\lambda_0^2} + \epsilon^2\sum_{j=1}^{\infty}\mathsf{E}\,(\boldsymbol{v}, \boldsymbol{H}_{j+1}\boldsymbol{q}^{j-1}\boldsymbol{H}_1\boldsymbol{u}) \\
&\quad - \epsilon^2\mathsf{E}\,(\boldsymbol{v}, \boldsymbol{H}_2\boldsymbol{u})(\boldsymbol{v}, \boldsymbol{H}_1\boldsymbol{u}).
\end{aligned} \tag{12.1.16}$$

2 Serially Independent Random Variation

Focus now on the important case of I.I.D. fluctuations in the vital rates. I will drop the parameter ϵ and simply refer to $(\epsilon\boldsymbol{H}_t)$ as the deviation. In the absence of serial autocorrelation only the first two terms of (12.1.16) survive and

$$a \simeq \log\lambda_0 - \frac{(\boldsymbol{v}\otimes\boldsymbol{v})^T\boldsymbol{c}(0)(\boldsymbol{u}\otimes\boldsymbol{u})}{2\lambda_0^2}. \tag{12.2.1}$$

As one expects, $a < \log \lambda_0$ when $\boldsymbol{c}(0) \neq 0$. It is very informative to rewrite (12.2.1) by making use of the sensitivity analysis of Section 11.1. Recall from there that for the average matrix $\boldsymbol{b}$ one has

$$\frac{\partial \lambda_0}{\partial b_{ij}} = v(i)u(j). \tag{12.2.2}$$

(Here $\boldsymbol{v}^T \boldsymbol{u} = 1$ by suitable normalization). The reader may now deduce that (12.2.1) can be rewritten as

$$a \simeq \log \lambda_0 - \frac{1}{2\lambda_0^2} \sum_{(ij)(k\ell)} \left(\frac{\partial \lambda_0}{\partial b_{ij}}\right) \left(\frac{\partial \lambda_0}{\partial b_{k\ell}}\right) \mathrm{Cov}\,(ij, k\ell), \tag{12.2.3}$$

where the sum is over $1 \leq i, j, k, \ell \leq n$ and

$$\mathrm{Cov}\,(ij, k\ell) = \mathsf{E}\,(\boldsymbol{H}_t)_{ij}(\boldsymbol{H}_t)_{k\ell}. \tag{12.2.4}$$

The sensitivity of a to random variation is therefore measured by squares (and pairwise products) of the sensitivity of λ_0 to average rates. This fact determines the differing effects of random variation in different vital rates.

3 Serially Independent Variation in Age-Structured Populations

Age-structured populations are the motivation for this theory and their projection matrices are Leslie matrices as in Section 2.4. Using the notation of that section, the stochastic growth rate for serially independent fluctuations in all vital rates is

$$\begin{aligned} a \cong \log \lambda_0 &- \frac{1}{2\lambda_0^2} \sum_i \left(\frac{\partial \lambda_0}{\partial f_i}\right)^2 \mathrm{Var}\,(F_i) \\ &- \frac{1}{2\lambda_0^2} \sum_i \left(\frac{\partial \lambda_0}{\partial p_i}\right)^2 \mathrm{Var}\,(P_i) - \frac{1}{\lambda_0^2} \sum_{i \neq j} \left(\frac{\partial \lambda_0}{\partial f_i}\right) \left(\frac{\partial \lambda_0}{\partial f_j}\right) \mathrm{Cov}\,(F_i, F_j) \\ &- \frac{1}{\lambda_0^2} \sum_{i \neq j} \left(\frac{\partial \lambda_0}{\partial p_i}\right) \left(\frac{\partial \lambda_0}{\partial p_j}\right) \mathrm{Cov}\,(P_i P_j) \\ &- \frac{1}{\lambda_0^2} \sum_{i \neq j} \left(\frac{\partial \lambda_0}{\partial p_i}\right) \left(\frac{\partial \lambda_0}{\partial f_j}\right) \mathrm{Cov}\,(P_i F_j). \end{aligned} \tag{12.3.1}$$

The variance and covariance notation should be transparent. *E.g.*, $\mathrm{Var}(F_i)$ is the variance in fertility at age i while $\mathrm{Cov}(P_i F_j)$ is the covariance between survival rate at age i and fertility at age j.

The results of Sections 2.4 and 11.1 can be combined to obtain important special cases of (12.3.1). I will list the final result in each case and let the reader do her own algebra.

1. *Fluctuations in juvenile survival,* taken to mean survival of the youngest age class. If this survival rate is a random $P_{1t} = p_1 + Z_t$, $E(Z_t) = 0$, $\mathrm{Var}(Z_t) = \sigma^2 = c^2 p_1^2$, then

$$a \cong \log \lambda_0 - \frac{c^2}{2T_0^2}\left(1 - \frac{f_1}{\lambda_0}\right)^2. \qquad (12.3.2)$$

2. *Uncorrelated fluctuations in all fertilities.* Thus at each age i the fertility at time t is $F_{i,t} = f_i + Z_{it}$, $\mathsf{E}(Z_{it}) = 0$, $\mathrm{Var}(Z_{it}) = \sigma_i^2$, $\mathrm{Cov}(Z_{it}\, Z_{jt}) = 0$ for $i \neq j$. Here

$$a \cong \log \lambda_0 - \frac{\sum_i (\ell_i \lambda_0^{-i} \sigma_i)^2}{2T_0^2}. \qquad (12.3.3)$$

3. *Correlated fertility fluctuations.* Here we take $F_{it} = f_i + Z_{it}$ with $E(Z_{it}) = 0$, $\mathrm{Cov}(Z_{it}\, Z_{jt}) = \sigma_i \sigma_j$, so that

$$a \cong \log \lambda_0 - \frac{\left(\sum_i \ell_i \lambda_0^{-i} \sigma_i\right)^2}{2T_0^2}. \qquad (12.3.4)$$

There are many important applications of these results, some of which are described in Chapters 14–17 of this book.

4 Higher-Order Expansions and the Loss of Ergodicity

The formulation of Section 12.1 clearly can be extended to higher orders by carrying the expansion (12.1.6) to higher orders in ϵ. The results are instructive although the algebra is not. (Details of the analysis will appear in a forthcoming paper). Here I present the results of a 4th order expansion for the case of I.I.D. random variation in the transition rate from class 1 to class 2. In the age-structured case, this is just the survival rate.

The expansion method through order ϵ^4 yields the result

$$\begin{aligned} a \cong \log \lambda_0 &- \frac{\epsilon^2 \theta^2 \sigma^2}{2\lambda_0^2} + \frac{\epsilon^3 \theta^3 \mu_3}{3\lambda_0^3} - \frac{\epsilon^4 \theta^4 \mu_4}{4\lambda_0^4} \\ &+ \frac{\epsilon^4 \theta^3 \sigma^4}{\lambda_0^4} \boldsymbol{w}^T (1 - \boldsymbol{q})^{-1} \boldsymbol{y} \\ &- \frac{\epsilon^4 \theta^2 \sigma^4}{2\lambda_0^4} (\boldsymbol{w} \otimes \boldsymbol{w})^T (1 - \boldsymbol{q} \otimes \boldsymbol{q})^{-1} (\boldsymbol{y} \otimes \boldsymbol{y}). \qquad (12.4.1) \end{aligned}$$

A novel feature of this expansion is the appearance of the inverse matrices shown. Recall that $\boldsymbol{q}$ has eigenvalues $(\lambda_\alpha / \lambda_0)$ so $\boldsymbol{q} \otimes \boldsymbol{q}$ has eigenvalues

$(\lambda_\alpha \lambda_\beta / \lambda_0^2)$ for $\alpha \neq 0$, $\beta \neq 0$. Therefore the expansion (12.4.1) converges only when $\lambda_0 > |\lambda_\alpha|$ for all $\alpha \neq 0$.

When demographic weak ergodicity is lost, the matrix $\boldsymbol{b}$ fails to be primitive and this corresponds to having one or more λ_α such that $|\lambda_\alpha| = \lambda_0$. The expansion (12.4.1) is therefore *divergent* when demographic ergodicity is lost. We see here the genesis of the singular behavior of a near parameter values where demographic ergodicity fails. Compare the discussion of the semelparous limit in Section 8.2.4.

5 Other Exponents

In Section 4.2 we considered the other Liapunov characteristic exponents of the random matrix products. These can also be analyzed by the expansion method described here; see Tuljapurkar (1989) for an example. Different applications of the technique are described in the next chapter.

13

POPULATION STRUCTURE FOR SMALL NOISE

This chapter extends the expansion method of Chapter 12 to the population structure vector $\boldsymbol{Y}_t$ and the reproductive value vector $\boldsymbol{V}_t$. As a byproduct of this analysis we get information on the growth rate of population over time, and on the serial correlation structure of populations over time in a varying environment. We learn how the history of environmental perturbations is filtered by population response. The first section below deals with the method itself while later sections consider implications of the results.

1 Expansions for Structure and Reproductive Value

The decomposition (12.1.1) and its attendant features from Section 12.1 are the basis for what we do here. The population structure vector $\boldsymbol{Y}_t$ obeys the equation

$$\boldsymbol{Y}_t = \boldsymbol{M}_t\boldsymbol{Y}_0/(\boldsymbol{e},\boldsymbol{M}_t\boldsymbol{Y}_0). \tag{13.1.1}$$

where the matrix $\boldsymbol{M}_t$ is the (now familiar) product of random matrices

$$\boldsymbol{M}_t = \boldsymbol{X}_t\boldsymbol{X}_{t-1}\ldots\boldsymbol{X}_1. \tag{13.1.2}$$

Similarly the (normalized, *i.e.*, components sum to unity) reproductive value vector obeys the backward-in-time equation

$$\boldsymbol{V}_1 = \boldsymbol{M}_t^T\boldsymbol{V}_{t+1}/(\boldsymbol{e},\boldsymbol{M}_t^T\boldsymbol{V}_{t+1}). \tag{13.1.3}$$

The objective here is to use these equations to gain insight into the statistical properties of $\boldsymbol{Y}$ and $\boldsymbol{V}$. Accordingly I focus on the long-time behavior of these vectors, when the effects of initial conditions have been erased by time. Assuming that demographic weak ergodicity holds for (13.1.1), and therefore also for (13.1.3), we may choose an initial vector that makes calculations convenient, without having to worry about the effects of the particular choice made (at least for steady-state statistical properties).

The choices we make are

$$\boldsymbol{Y}_0 = \boldsymbol{u}, \tag{13.1.4}$$

and

$$\boldsymbol{V}_{t+1} = \boldsymbol{v}. \tag{13.1.5}$$

The next step is to use the expansion (12.1.7) in (13.1.1, 3) along with the choices above, and obtain the corresponding expansions of $\boldsymbol{Y}_t$ and $\boldsymbol{V}_1$ to second-order in the parameter ϵ (which measures the amplitude of environmental fluctuations). The results are:

$$\boldsymbol{Y}_t \simeq \boldsymbol{u} + \epsilon(1 - \boldsymbol{u}\boldsymbol{e}^T)\boldsymbol{R}_{1t} + \epsilon^2(1 - \boldsymbol{u}\boldsymbol{e}^T)\left[\boldsymbol{R}_{2t} - (\boldsymbol{e}, \boldsymbol{R}_{1t})\boldsymbol{R}_{1t}\right], \quad (13.1.6)$$

where

$$\boldsymbol{R}_{it} = \boldsymbol{S}_{it}\boldsymbol{u}/\lambda_0^t, \quad i = 1, 2. \quad (13.1.7)$$

Similarly,

$$\boldsymbol{V}_1 \simeq \boldsymbol{v} + \epsilon(1 - \boldsymbol{v}\boldsymbol{e}^T)\boldsymbol{P}_{1t} + \epsilon^2(1 - \boldsymbol{v}\boldsymbol{e}^T)\left[\boldsymbol{P}_{2t} - (\boldsymbol{e}, \boldsymbol{P}_{1t})\boldsymbol{P}_{1t}\right], \quad (13.1.8)$$

where

$$\boldsymbol{P}_{it} = \boldsymbol{S}_{it}^T\boldsymbol{v}/\lambda_0^t, \quad i = 1, 2. \quad (13.1.9)$$

In working further with these expressions it is useful to note some properties of the objects $(1 - \boldsymbol{u}\boldsymbol{e}^T) = \boldsymbol{k}$, say, and $(1 - \boldsymbol{v}\boldsymbol{e}^T) = \boldsymbol{h}$. First, direct calculation shows that

$$\boldsymbol{k}^2 = \boldsymbol{k}, and \qquad \boldsymbol{h}^2 = \boldsymbol{h}. \quad (13.1.10)$$

Further, if we apply matrix $\boldsymbol{k}$ to any power of the average vital rate matrix, then the spectral decomposition (see (12.1.14)) shows that

$$\begin{aligned} \boldsymbol{k}\boldsymbol{b}^m &= \lambda_0^m \boldsymbol{k}(\boldsymbol{u}\boldsymbol{v}^T + \boldsymbol{q}^m) \\ &= \lambda_0^m \boldsymbol{k}\boldsymbol{q}^m, \quad m \geq 1. \end{aligned} \quad (13.1.11)$$

Similarly, applying $\boldsymbol{h}$ to the transpose of any power of $\boldsymbol{b}$ shows that

$$\boldsymbol{h}(\boldsymbol{b}^T)^m = \lambda_0^m \boldsymbol{h}(\boldsymbol{q}^T)^m, \quad m \geq 1. \quad (13.1.12)$$

Finally, note that

$$(\boldsymbol{e}, \boldsymbol{k}) = (\boldsymbol{e}, \boldsymbol{h}) = 0 \quad (13.1.13)$$

This last fact is the least significant, and it simply ensures that the vectors $\boldsymbol{Y}$ and $\boldsymbol{V}$ in the expansions have the right normalization. However (13.1.10–12) show that $\boldsymbol{k}$ and $\boldsymbol{h}$ act as projection operators which extract the transient component (remember that $\boldsymbol{q}^m \to 0$ as m increases) from powers of the average vital rate matrix.

2 Properties of Structure and Reproductive Value

The expansion (13.1.7) reveals several facts about $\boldsymbol{Y}_t$. Recall from the definitions in Section 12.1 that $\mathsf{E}(\boldsymbol{R}_{it}) = 0$, since that term is linear in the random deviations $\boldsymbol{H}_t$. Therefore the average value of $\boldsymbol{Y}_t$ differs from

$\boldsymbol{u}$ (which, remember, is the stable structure corresponding to the average vital rates) by a term of magnitude ϵ^2. The current population structure, $\boldsymbol{Y}_t$, however, deviates from $\boldsymbol{u}$ by a term of order ϵ, *i.e*, by an amount which can be considerably larger than the difference between $\mathsf{E}(\boldsymbol{Y}_t)$ and $\boldsymbol{u}$. The term involving $\boldsymbol{R}_{1t}$ is a weighted average of recent environmental "shocks" to the vital rates, and the most recent shocks are most heavily weighted. To see why, use (13.1.12) in (13.1.8) with (12.1.7) to find that

$$(1 - \boldsymbol{u}\boldsymbol{e}^T)\boldsymbol{R}_{1t} = (1 - \boldsymbol{u}\boldsymbol{e}^T)\left(\boldsymbol{H}_t + \boldsymbol{q}\boldsymbol{H}_{t-1} + \boldsymbol{q}^2\boldsymbol{H}_{t-2} + \ldots\right)\boldsymbol{u}. \quad (13.2.1)$$

Since $\boldsymbol{q}^m \to 0$ as m increases, terms involving random deviations from the distant past are considerably less important here than more recent deviations.

The variance of $\boldsymbol{Y}_t$ may be computed to second order by using (13.1.7) to compute the variance and then taking the limit $t \to \infty$. I will not present this calculation here but will turn to the more interesting, and more general, computation of the serial correlation between population structure at two times, say t and $t+m$, for some positive integer m. To order ϵ^2, it follows directly from (13.1.7) that

$$\begin{aligned} \boldsymbol{c}_y(m) &= \mathsf{E}(\boldsymbol{Y}_{t+m} - \mathsf{E}\boldsymbol{Y}_{t+m})(\boldsymbol{Y}_t - \mathsf{E}Y_t) \\ &= \epsilon^2 \boldsymbol{k} \otimes \boldsymbol{k}\mathsf{E}(\boldsymbol{R}_{1,t+m} \otimes \boldsymbol{R}_{1t}). \end{aligned} \quad (13.2.2)$$

The terms in (13.2.2) are complicated and it is illuminating to start by examining them in the simplest case. When there is *no* serial autocorrelation in the random vital rates, there *is* nonetheless serial autocorrelation of population structure, and we have

$$\boldsymbol{c}_y(m) = \epsilon^2(1 - \boldsymbol{k}\boldsymbol{q} \otimes \boldsymbol{k}\boldsymbol{q})^{-1}\mathsf{E}(\boldsymbol{q}^m\boldsymbol{H}_1 \otimes \boldsymbol{H}_1)(\boldsymbol{u} \otimes \boldsymbol{u}). \quad (13.2.3)$$

Observe that this serial correlation falls off at the characteristic rate of population transients which are determined by $\boldsymbol{q}$ (and thus by average vital rates). Thus, this serial correlation can be quite long-lived even when vital rates change in a purely random uncorrelated way from one time interval to the next.

When there is serial autocorrelation in the vital rates over time, one can use a simple device to obtain the power spectrum of fluctuations in the population structure, avoiding the tedium of expanding (13.2.2) term-by-term. The power spectrum is the discrete Fourier transform of the expression (13.2.2) for the serial autocorrelation, and may be found by the indirect method of first determining the Fourier transform of the series $\boldsymbol{k}\{\boldsymbol{Y}_t - \mathsf{E}(\boldsymbol{Y}_t)\}$. Using (13.1.7) and working to order ϵ, note that this transform in turn requires the Fourier transform of $\boldsymbol{k}\boldsymbol{R}_{1t}$. Inspection of (13.2.1) shows that the latter quantity has the structure of a discrete convolution, and we immediately have

$$\sum_t e^{-i\omega t}\boldsymbol{k}\boldsymbol{R}_{1t} = \boldsymbol{k}\pi_q(\omega)\pi_H(\omega), \quad (13.2.4)$$

where

$$\pi_q(\omega) = \sum_j e^{-i\omega j} \boldsymbol{q}^j, \quad \pi_H(\omega) = \sum_j e^{-i\omega j} \boldsymbol{H}_j. \tag{13.2.5}$$

The power spectrum of fluctuations in $\boldsymbol{Y}_t$ is therefore given by the object

$$\pi_y(\omega) = (\boldsymbol{k} \otimes \boldsymbol{k}) \left(\pi_q(\omega) \otimes \pi_q^*(\omega)\right) \mathsf{E} \left(\pi_H(\omega) \otimes \pi_H^*(\omega)\right), \tag{13.2.6}$$

where the asterisks indicate complex conjugates. The right-hand-side of this equation is the product of the power spectrum of transients (determined by $\boldsymbol{q}$) and of the fluctuations. It follows that the transient oscillations governed by average rates are modulated by the amplitude of serial autocorrelations in the rates at corresponding periods. Similar qualitative and quantitative properties of the reproductive value vector may be deduced from (13.1.9).

3 Applications

The results given here are have great potential for application. In human population projection they make possible extensions of the work of Lee (1974), who constructed a serially-autocorrelated model for human fertility. The objective was to combine the stochastic features of the rates with average vital rates to predict statistical features of population age-structure. I consider such work in more detail in the next chapter, but there is great scope for analyses which exploit the methods of this chapter.

A related application is in analyzing the phenomenon of dominant cohorts, which is well known in fisheries and also other ecological systems. The essence of the phenomenon is that series of population observations reveal that population composition is dominated over many time-periods by a single cohort which eventually dies out, only to be followed some time later by another dominant cohort. Thus the population structure displays a pattern of serial correlation, and it would be useful to relate this to underlying patterns in the environment. It should be obvious that the results given above are tailor-made for an exploration of this phenomenon, but the subject is unexplored.

A final interesting application is to the temporal pattern of variability in one-period population growth rates. Since both vital rates and population structure fluctuate over time, so will the one-period growth rates. If one has historical data on these growth rates, will they be of any value in predicting the future? This question is particularly relevant to harvested populations, and to populations which are threatened by extinction. Recall that the growth rate of population between times $t-1$ and t is given by

$$\lambda_t = (\boldsymbol{e}, \boldsymbol{X}_t \boldsymbol{Y}_{t-1}) \tag{13.3.1}$$

Since we have detailed information on the vector $\boldsymbol{Y}_t$ in the steady-state (from Sections 1,2 above) it is possible to obtain the serial correlation of the series of growth rates to order ϵ^2.

14

POPULATION PROJECTION

Projection (or forecasting) is the first of the applications of the preceding theory to be discussed in this book. In discussing applications I aim to highlight the central issues and the most promising approaches for dealing with them. I will however be very selective.

Projection and estimation are complementary. To make projections we need to set confidence intervals of some kind around a point projection; we also need estimates of model parameters. The central issue is to determine how variance in vital rates works its way into population variance. Two approaches are considered here: one focuses on the long-run and obtains estimates of a and σ^2 from the data; the other makes use of data on varying vital rates and considers more detailed forecasts in the short-run.

1 Long-run Projections

The random-rates model leads asymptotically to simple exponential growth, as described by the lognormal limit theorem of Section 4.2.3. If we take the model's assumptions to be consistent with a set of historical data, then we can make projections by estimating the parameters in the limiting distribution of population number.

Heyde and Cohen (1985) provide estimators of a and σ^2 which make use of a time-series of counts of total population. Their theory was developed for a closed population but will work for at least some immigration patterns (Heyde 1985). The estimator for growth rate is the natural one,

$$\widehat{a} = (\log M_T - \log M_1)/(T-1) \qquad (14.1.1)$$

where the M's are total population counts over a span of T time units.

Their variance estimator is more complicated since they have to contend with the presence of serial autocorrelation in the M series resulting from multiplicative growth. I will present a little background to their main result. Consider the following general situation: we have a sequence $\{Z_n\}$ which is stationary, uniformly mixing, and has $\mathsf{E}(Z)=0$, $\mathsf{E}(Z^2)<\infty$. Let $S_n = \sum_{j=1}^{n} Z_j$, and suppose that

$$\mathsf{E}\, S_n^2/n \to \sigma_1^2, \qquad \mathsf{E}\,|S_n|/\sqrt{n} \to \sigma_1(2/\pi)^{\frac{1}{2}}, \quad \text{as } n\to\infty.$$

Heyde and Cohen establish that if we choose a set of positive constants $\{c_i\}$ and set $b_n = \sum_{i \le n} c_i$, then a consistent estimator of σ_1 is the object

$$(1/b_n) \sum_{i=1}^{n} c_i |S_i| / \sqrt{i}$$

There are certain conditions on the magnitudes of the $\{c_i\}$ relative to the $\{b_n\}$ which are satisfied by the particular choice $c_i = (1/i)$. To apply this result to the population case, Heyde and Cohen show that

$$\lim_{t \to \infty} \frac{1}{\sqrt{t}} \mathsf{E}\, |\log M_t - at| = \sigma(2/\pi)^{\frac{1}{2}} \tag{14.1.2}$$

They then apply the estimator described above, taking $c_i = (1/i), b = \log i$ and show that it is indeed consistent for the variance. Hence their basic variance estimator for a time series of T counts starting at time $t_0 + 1$ is

$$D(T, t_0) = (\pi/2)^{\frac{1}{2}} \, [1/\log(T-1)] \sum_{j=1}^{T-1} j^{-3/2} \, |\log M_{t_0+j+1} - \log M_{t_0+1} - j\widehat{a}| \, . \tag{14.1.3}$$

where $\widehat{a}$ is the mean estimated for the same data. They present some numerical experiments and argue that for finite samples the smallest standard deviation of the estimate is obtained with a linear combination of two estimates of the form (14.1.3) which are displaced by one time interval at each end,

$$\widehat{\sigma} = \frac{1}{2} \, [D(T, t_0) + D(T-1, t_0+1)] \tag{14.1.4}$$

Given the estimate (14.1.4) one can put $100\alpha\%$ confidence intervals on the growth rate a of the form

$$\widehat{a} \pm z(\alpha)(\widehat{\sigma}/\sqrt{t})$$

where $z(\alpha)$ is a positive number such that the area to the right of it under a standard normal curve is $(1-\alpha)/2$. In addition the asymptotic independence of total population number at widely separated times can be used to construct confidence intervals for forecasts at future times given a finite data set.

Heyde and Cohen give an illustration of the use of their estimator in their paper. A further application is described by Cohen (1986) who considers a sequence of 41 estimated total population sizes for Sweden at 5-year intervals from 1780 to 1980 inclusive. He used the estimator (14.1.4) on subsets of these data to make projections and confidence intervals at various times forward from the end of each data subset. He compares the confidence intervals so obtained with intervals based on a forecast error model due to Williams and Goodman (1971) and also a pair of empirically derived estimators suggested by Stoto (1983). He concludes that the first two confidence intervals should generally be in close agreement, and that Stoto's optimistic estimate is a reasonable one for developed countries. He points out that many open questions remain.

2 Short-run Projection

In many practical situations one would like to make the maximum use of the population information available to us, which is not true for the long-run method of the previous section. Ronald Lee (personal communication) has pointed out that long-run estimators of a and σ^2 ignore all age-structure information and are probably not optimal in the short and medium term. How does one do better?

Lee (1974) used the procedure of fitting a time series model to fertility rates and then generated population forecasts and confidence intervals based on these rates. His paper resulted in considerable further work which exploited the well-known time series methodology of Box and Jenkins (1970) to examine a variety of ARMA (autoregressive-moving average) models (*e.g.*, MacDonald 1979, Saboia 1977, Lee 1977). A related but potentially more general approach is taken by Alho and Spencer (1985). Tuljapurkar (1987) has studied the more general problem of combining a stochastic model for vital rates with the matrix projection model. I will discuss here the limitations of the earlier methods and the directions in which that work may be extended, as well as a general feature of the combined model.

Lee (1974) and many subsequent workers have started with what I will call a time-series-matrix (TSM) projection model. In this model the vital rates are represented by a linear time-series model and their updated values are inserted into the projection matrix. Lee simplified the resulting random-rates model by linearizing the equation, based on the following assumptions and arguments. First, the emphasis is on populations near stationarity so that the average net reproductive rate is taken to be unity. Stochastic fluctuations in the net rate are supposed to be small, and it is assumed that the average population structure will be equal to the stable age-structure $\boldsymbol{u}$ that would obtain if the vital rates were fixed at their average values. The total population vector can then be described as proportional to a sum of the type $(\boldsymbol{u} + \boldsymbol{G}_t)$, where $\boldsymbol{G}_t$ is a small deviation whose average does not grow over time. The vital rates are also decomposed in the manner of Section 12.1, and the two decompositions are inserted into the projection equation. The result is a linear stochastic difference equation of the form

$$\boldsymbol{G}_{t+1} = \boldsymbol{b}\boldsymbol{G}_t + \epsilon \boldsymbol{H}_{t+1}\boldsymbol{u} \tag{14.2.1}$$

The theory of linear stochastic equations informs us that (1) the variance of $\boldsymbol{G}_t$ and thus of population size will increase linearly with time; (2) the linear rate at which variance increases will become larger when the random terms exhibit (usually) positive serial autocorrelation; (3) the vector $\boldsymbol{G}_t$ describes the transients in population structure, and the autocorrelation structure of these transients is determined in large part by the transients in the average vital rate matrix $\boldsymbol{b}$. This means that the dominant periodicity excited by small stochastic variability will be roughly a generational cycle. All of these features are discussed (less telegraphically) by Lee.

Tuljapurkar (1987) considers the TSM model without linearization, in which case the main results of the small-noise expansions in Chapters 12 and 13 can be applied directly. The main conclusions are:

1. assuming the average net reproductive rate to be unity does not guarantee that a is zero, and for the patterns of serial autocorrelation expected of human fertility the value of a is actually less than zero. The value of a decreases quadratically in the amplitude (ϵ) of the fluctuations in fertility.

2. the average population structure (averaged over long times) is not very different from $\boldsymbol{u}$, with proportions in the two vectors being within about 5% of each other.

3. over any finite period the actual population structure varies considerably over time, and differs very markedly from $\boldsymbol{u}$.

4. the pattern of covariation of age-groups is strongly dependent on the temporal structure of vital rates, with characteristics that can be understood using the methods of Chapter 13.

Finally I will discuss a computation that yields considerable insight into the relative importance of vital rate fluctuations and age structure fluctuations. Lee's 1974 paper asserted that the variance in population growth rates should be dominated by the variance in vital rates, as opposed to the variance in population structure. Tuljapurkar and Lee (1987, unpublished) used the methods of Chapter 13 to compute the relative contributions, as follows. From (13.3.1) and the decomposition (12.1.1), the one-period growth rate is

$$\begin{aligned} \lambda_t &= (\boldsymbol{e}, (\boldsymbol{b} + \epsilon \boldsymbol{H}_t)(\boldsymbol{u} + \boldsymbol{Y}_t - \boldsymbol{u})), \\ &= (\boldsymbol{e}, \boldsymbol{b}\boldsymbol{u}) + \epsilon(\boldsymbol{e}, \boldsymbol{H}_t \boldsymbol{u}) + (\boldsymbol{e}, \boldsymbol{b}(\boldsymbol{Y}_t - \boldsymbol{u})) + \epsilon(\boldsymbol{e}, \boldsymbol{H}_t(\boldsymbol{Y}_t = \boldsymbol{u})), \\ &= \lambda_0 + \epsilon \Delta_R + \Delta_A + \epsilon \Delta_{AR} \end{aligned} \tag{14.2.2}$$

The two terms of order ϵ in this equation can be interpreted as follows: Δ_R is the average vital rates acting on the fluctuations in population structure, while Δ_A is the fluctuations in vital rates acting on average population structure. These identifications ignore the interaction term but are close enough for our purposes. We can now define the variance in growth rate due to variance in vital rates as

$$V_R = \mathsf{E}\,(\Delta_R^2), \tag{14.2.3}$$

and the variance in growth rate due to variance in population structure as

$$V_A = \mathsf{E}\,(\Delta_A^2) \tag{14.2.4}$$

The value in (14.2.3) is easily computed by the methods of Chapter 12, while (13.1.6) can be used to compute the right-side of (14.2.4). The result can then be used with any set of vital rates to compute the ratio V_R/V_A. For rates similar to those of the US population in 1960, the analytical result yields a ratio of approximately 0.7; this value is confirmed by simulations. Thus about one-third of the variance in growth rate is due to fluctuations in age-structure.

15

LIFE HISTORY AND ITEROPARITY

1 The Problems

Fisher (1930) posed the central question about life histories, asking how the apportionment of reproduction over life might have evolved. Cole (1954) framed the style of many recent studies in comparing the evolutionary advantage of semelparity (reproducing once) and iteroparity (reproducing more than once). Since then there has been considerable work on general features of life history evolution (*e.g.*, Williams 1966, the review by Stearns 1976, Begon, Harper and Townsend 1986), and on features specific to certain species or genera (*e.g.*, Denno and Dingle 1981, Jackson, Buss and Cook 1985). It is clear from Lewontin (1965) that classical demographic arguments can only account for some of the life historical patterns in nature, and later workers have tried to incorporate factors outside the classical framework. One of these is an environment which produces random variations in vital rates and thereby generates selection pressures on life histories.

Although there has been considerable interest in the evolutionary consequences of random vital rates, the theoretical picture has been rather confusing. The review by Stearns (1976) features two contradictory views of how randomness affects life history evolution. One, the r-K view, relies on deterministic theory to conclude that so-called "r-selected" life histories are most advantageous in the presence of variability. The other, the bet-hedging view, uses some version of stochastic theory to argue that "more" iteroparous life histories are more advantageous. This tentative state of the theory undoubtedly is due to the fact that until recently no analysis has dealt with (1) age structure, (2) random environmental variability, (3) life histories of arbitrary length, (4) variation and covariation of components of the life history, (5) a genetic basis for life history differences, to predict the rate and direction of natural selection on particular types of life histories.

I will discuss here the results of such analyses which use the theory developed in this book. The genetic basis is the ESS criterion for allelic invasion discussed in Chapter 6. Since I will deal only with age-structured populations, the fitness measure of interest is the stochastic growth rate a. Thus the present theory has the same genetic underpinning as the classical theory for unchanging environments (*cf.* Charlesworth 1980). In practical terms, the central task is to understand how a depends on the components of the life history phenotype. This is a difficult question but considerable progress has been made in answering it.

2 The Crossover Effect

The central biological theme of this chapter is that random variation in vital rates causes *major* departures from the views of life history evolution which are based on fixed rates and stable age structures. To illustrate this theme, consider a set of alternative life histories. Note that a life history in a random environment is described by the average vital rates *plus* the random deviations from the averages which occur over time. In the absence of random variation, the average rates determine the classical Malthusian parameter r_0 which is a measure of the relative fitness of the underlying genotype. In a random environment, one must compare a values between phenotypes. The crucial new feature that emerges is that random variation in rates can completely reverse the relative fitness ranking of life history phenotypes; I call this the **crossover effect**.

An illustration is provided by a set of life histories in which random variation affects the survival of the youngest age class. Consider three life histories (in the spirit of Murphy 1968) which have geometrically declining survivorship curves, equal fertilities in all reproductive years (Figure 15.2.1) and the same average net reproduction rate R_0. Suppose that the survival rate of juveniles in these life histories is variable; this translates into random variation of adult fertility with perfect correlation between all pairs of fertilities. If the coefficient of variation of the fertilities is c, the analytical approximation to a (*cf.* Chapter 12) is

$$a = r_0 - \frac{c^2}{2T_0^2}. \tag{15.2.1}$$

Here r_0 and T_0 are the long run growth rate and mean length of generation (see Chapter 2 for definitions) determined by the average vital rates, and we assume reproduction begins at age 2 or later. Equation (15.2.1) implies that as c (the level of uncertainty in the environment) increases, the value of a will decrease. In addition, the decrease will be relatively greater for populations with lower mean generation lengths. The 3 life histories in Figure 15.2.1 are labelled so that $r_0(1) > r_2(2) > r_0(3)$. As the figure suggests, the mean lengths of generation are ordered as $T_0(1) < T_0(2) < T_0(3)$. From (15.2.1) we can therefore predict that for sufficiently high c there will be a *crossover* in growth rates, so that $a(1) < a(2) < a(3)$. This analytical prediction is nicely borne out by simulations with the result shown in Figure 15.2.2: sufficiently high variability in the environment reverses the relative advantages of these three life histories.

This example shows that uncertainty can have dramatic impact. However, it is premature to conclude that iteroparity is always advantageous in

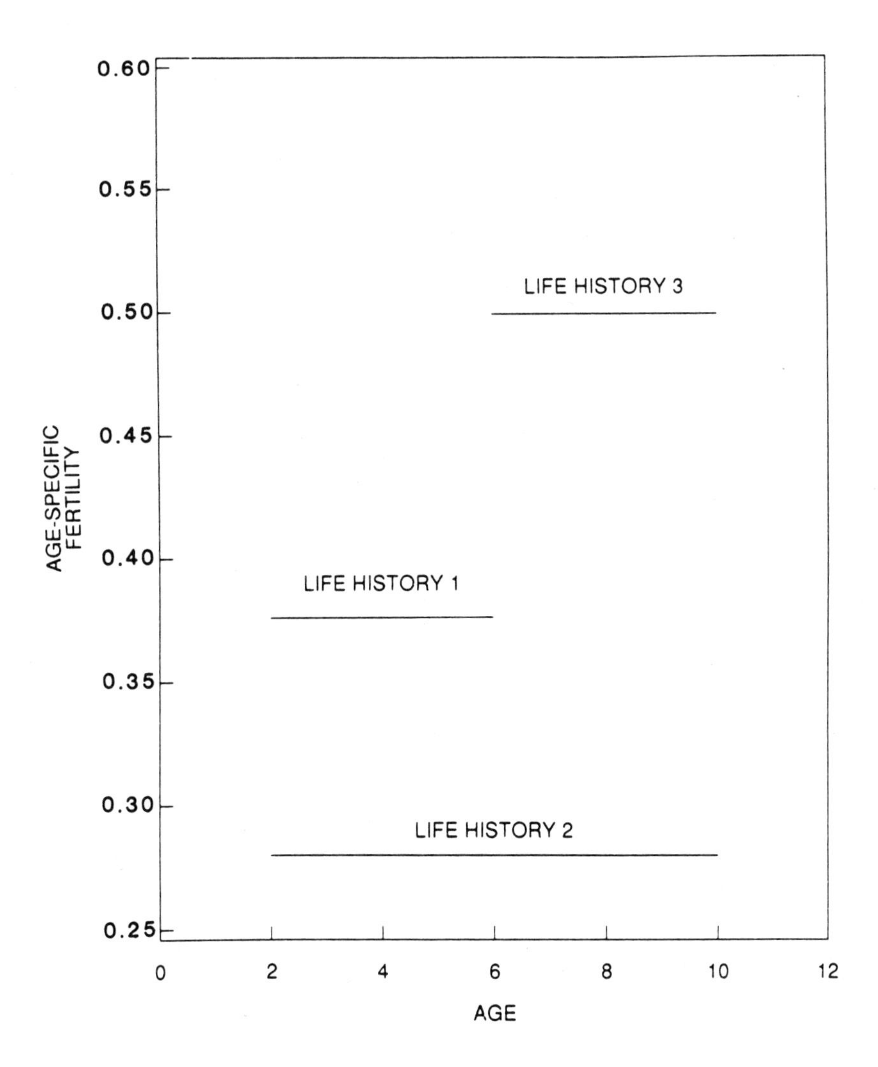

FIGURE 15.2.1. Three stylized life histories with equal net reproductive rates and deterministic growth rates ranked as $r_1 > r_2 > r_3$

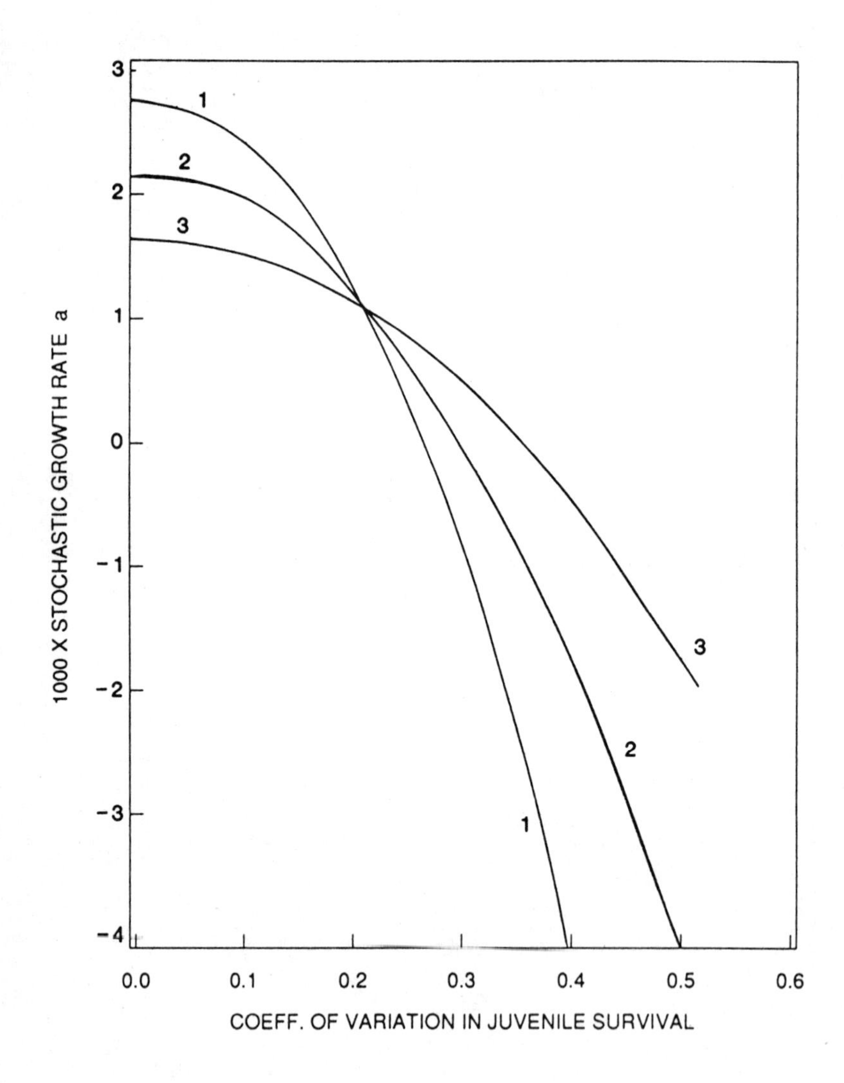

FIGURE 15.2.2. Stochastic growth rate for increasing variation for the three life histories of the preceding figure. For high variability the ranking of stochastic growth rates is the reverse of the ranking of deterministic growth rates

a randomly varying environment. The magnitude of random variability and the life history structure are important in the ranking of life histories, and distinct life histories can have very similar a for the same level of random variation.

3 Iteroparity

The problem: does environmentally driven fluctuation in vital rates result in an evolutionary advantage for iteroparous life histories? This question was addressed by Murphy (1968), Schaffer (1974b), Giesel (1976), Hastings and Caswell (1979), and Goodman (1984), none of whom used the theory of stochastic demography. Orzack and Tuljapurkar (1989) used this theory, taking iteroparity to be a continuous character described by the temporal clumping and positioning of reproduction during life. They used the analytical results of this book and simulation results to study the relative fitness, a, of iteroparous life histories chosen from constrained phenotype sets. I adopt this approach here to describe and extend their results.

The essence of iteroparity is the spreading out of reproduction over life. So it is reasonable to construct a spectrum of life histories in which reproduction is (1) concentrated early in life, or (2) spread out over increasing life spans, or (3) concentrated toward the end of life. In going from (1) to (3), we pass from early and nearly semelparous reproduction to increasing iteroparity and finally to late and nearly semelparous reproduction. In order to make meaningful comparisons, it is essential to constrain the set of life histories. I follow Orzack and Tuljapurkar (1989) and Murphy (1968), by supposing that average lifetime weighted reproduction ($=R_0=$ the net reproductive rate $=\sum_i \ell_i f_i$) is the same for all life histories. Other constraints can be chosen using, *e.g.*, total lifetime fertility, and do not appear to change the results as long as the constraint criterion is correlated with R_0.

The average vital rates for the life histories in this phenotype set guarantee that "early, concentrated" reproduction results have the highest r_0. This leads to the classical optimality argument for r–selected life histories. What happens with random rates? I first use analytical results to get the answer, and then summarize related numerical results.

3.1 Iteroparity is important—analytics

To use the analytical results of Chapter 12, we must specify the nature of random variation in vital rates. Focus on the relatively simple situation in which fertility at every age varies randomly with the same coefficient of variation, so that the (random) fertility of age class i at time t is

$$F_{i,t} = f_i + c f_i Z_t \qquad (15.3.1)$$

$$\mathsf{E}\, Z_t = 0, \quad \mathrm{Var}(Z_t) = 1. \tag{15.3.2}$$

Assume there is no serial autocorrelation. These assumptions correspond to two different biological situations: (i) the numbers of offspring produced actually varies as indicated, or (ii) offspring numbers are fixed but the survival of individuals from birth till they are counted in the first age class is random.

Given (15.3.1-2) the analytical approximation to the stochastic growth rate is

$$a \simeq r_0 - \frac{c^2}{2T_0^2}. \tag{15.3.3}$$

Here r_0 is the long run growth rate determined by the average vital rates, and the mean length of generation corresponding to the average rates (see equation (2.4.10)) is

$$T_0 = \sum_i e^{-ir} \ell_i f_i. \tag{15.3.4}$$

From the classical theory, it is known (Keyfitz 1968) that when $R_0 = \sum_i \ell_i f_i$ is not too different from unity,

$$r_0 \simeq \frac{\log R_0}{T_0} + \frac{(\log R_0)^2 s_\phi^2}{2T_0^3}. \tag{15.3.5}$$

Here s_ϕ is a measure of how concentrated reproduction is over age,

$$s_\phi^2 = \sum_i i^2(\phi_i/R_0) - \Big[\sum_i i(\phi_i/R_0)\Big]^2. \tag{15.3.6}$$

It is just the variance in age of net reproduction around the mean T_0.

Now consider a life history set with constrained R_0 in which T_0 increases as we move through the set. From (15.3.5) it is clear that as we go through the set the increase in T_0 leads to a decrease in r_0. Indeed such sets of life histories lie near a curve of the form $r_0 T_0 =$ (a constant). For *comparative purposes*, we can use T_0 to index successive life histories in the set.

In the presence of random variation, (15.3.3) applies and we can ask: what happens to a for different life histories as c increases? The answer is that, because of the crossover effect, life histories with large T_0 can have a higher a when c is large than those with small T_0. The amount of random variation needed to produce crossover for a given pair of life histories can be computed as follows. Let r_A, r_B be the long run growth rates and T_A, T_B the mean lengths of generation determined by the average vital rates of life histories A and B. Suppose that $r_A > r_B$ (and $T_A < T_B$). Then the coefficient of variation at which the two have equal a is

$$c = 2T_A T_B(r_A - r_B)/\big[(T_A + T_B)(T_B - T_A)\big]. \tag{15.3.7}$$

In order for the crossover to be biologically meaningful, it should happen for coefficients of variation which are within the range of observed values.

From the examples in Chapter 5, the amount of variation experienced by organisms differs a great deal between species, between animals *vs.* plants, and so on. But (15.3.7) allows one to quickly assess the relevant c beyond which random rates start to exert a selection pressure counter to that revealed by average rates alone.

A different question to ask is: if we fix a particular value of a, how do life histories compare in the amount of randomness they can tolerate at that growth rate? For fixed a rewrite (15.3.3) as

$$c = \sqrt{2(r-a)}\, T_0. \tag{15.3.8}$$

The locus of (c, T_0) values given by (15.3.8) for a fixed a is an **indifference curve**, along which life histories are selectively neutral. It is clear that if we start with life histories of small T_0, then c increases with T_0 initially. However, from (15.3.5) we have that $(dr_0/dT_0) < 0$ for life histories in the constrained set, and so the indifference curve may reach a peak and then fall. The condition for this is that $(dc/dT_0) = 0$ in (15.3.8), and from (15.3.5) and (15.3.8) it follows that the peak occurs for T_0 which solves

$$2aT_0^3 - 3(\log R_0)T_0^2 + (5/2)s_\phi^2(\log R_0)^2 = 0. \tag{15.3.9}$$

Of course this value of T_0 may not lie within the set of life histories. If it does, the indifference curve will have the form shown in Figure 15.3.10. The most important feature of such a shape is that for any value of c, there are usually *two rather different life histories which are equally fit.*

Let us now turn to the work of Orzack and Tuljapurkar (1989) which fleshes out these analytical arguments.

3.2 ITEROPARITY IS IMPORTANT—NUMERICS

The numerical results described here involved both computations based on (15.3.3) and its generalizations (below) as well as simulations to check the analytical predictions. These predictions were very accurate and I concentrate on biological results.

A. *Life history sets*

Average life history sets with constrained R_0 were constructed in three geometrics. Tables 15.3.1, 15.3.2 show the "flat", "declining", and "peaked" life histories, displaying average net maternity values with age. These adjectives refer to the shape of $\phi_i (= \ell_i f_i)$ *vs.* age i, and each set covers a range of ages of first (α) and last (ω) reproduction. Thus each set goes over the spectrum from "early, concentrated" through "spread-out" to "late, concentrated" reproduction.

The specific results which I discuss here are a subset of the rather large collection for 25 life histories (9 flat, 9 declining, 7 peaked) studied for

FLAT LIFE-HISTORY SET USED FOR ANALYTIC AND NUMERICAL RESULTS

Life History	ϕ_i	α	ω	$\ln \lambda_0$	T_0	D
1	0.525	1	2	0.0326	1.492	0.5001
2	0.2625	1	4	0.0196	2.475	0.2501
3	0.175	1	6	0.0140	3.459	0.1668
4	0.13125	1	8	0.0109	4.443	0.1251
5	0.105	1	10	0.0089	5.426	0.1001
6	0.13125	3	10	0.0075	6.460	0.1250
7	0.175	5	10	0.0065	7.481	0.1667
8	0.2625	7	10	0.0057	8.493	0.2500
9	0.525	9	10	0.0051	9.499	0.5000

TABLE 15.3.1. Columns show the average net fertility (ϕ_i) for all ages from age of first reproduction (α) to age of last reproduction (ω). The dominant eigenvalue of the average Leslie matrix is λ_0, the mean length of generation is T_0, and D measures the diversity of the life history

many patterns of random variability. I focus on life histories with "flat" ϕ_is because: (1) the qualitative features of interest are the same regardless of the shape of ϕ_is, (2) the goal is to highlight conclusions which appear robust to geometrical changes in the life history set, and (3) I can make coherent comparisons within the set. Accordingly, most of the results for declining and peaked life histories are omitted.

The calculations assume that the coefficient of variation, c, is the same for all ϕ_{it}, independent of age. There are few data available to guide one's choice in this regard. The results of Rose and Charlesworth (1981) indicate that the coefficient of variation of daily fecundity increases with age in lines of *Drosophila melanogaster* sampled from a laboratory population. Alternatively, the data of Pinero and Sarukhan (1982) suggest that coefficients of variation of survivorship decline with age in the palm, *Astrocaryum mexicanum*. Ignoring (purely for the sake of argument) that these data do not relate directly to ϕ_{it} values, they most likely indicate that one could find species in which c values decrease, remain constant, or increase with age. The choice of an age-invariant c value is convenient for analysis and most likely applicable to many organisms but not the only choice one could make.

Numerical calculation of stochastic growth rates requires a choice about the distribution from which random "environmental" deviates are to be sampled. In Orzack and Tuljapurkar's (1989) simulations, the underlying distribution is lognormal. Several reasons motivated this choice. The first is that a continuous distribution was deemed more biologically realistic than one with discrete states and its use avoids possible mathematical singularities (*e.g.*, Chapter 8). Second, this distribution applies naturally to ϕ_{it} values since it arises as the limiting distribution of products of independent or weakly correlated positive random variables (*e.g.*, Johnson and Kotz 1970, p. 113). Finally, it is known that a skewed environmental dis-

THE AVERAGE VALUES OF "NET" FERTILITY AT EACH AGE

LIFE HISTORY	AGE												
	1	2	3	4	5	6	7	8	9	10	$\ln \lambda_0$	T_0	D
Declining													
1	0.5355	0.5145									0.0328	1.482	0.5007
2	0.2940	0.2730	0.2520	0.2310							0.0204	2.375	0.2531
3	0.2275	0.2065	0.1855	0.1645	0.1435	0.1225					0.0156	3.107	0.1755
4	0.2048	0.1838	0.1628	0.1417	0.1208	0.0997	0.0787	0.0578			0.0134	3.599	0.1443
5	0.1995	0.1785	0.1575	0.1365	0.1155	0.0945	0.0735	0.0525	0.0315	0.0105	0.0128	3.780	0.1359
6	0.0	0.0	0.2048	0.1838	0.1628	0.1417	0.1208	0.0997	0.0787	0.0578	0.0087	5.621	0.1434
7	0.0	0.0	0.0	0.0	0.2275	0.2065	0.1855	0.1645	0.1435	0.1225	0.0068	7.131	0.1745
8	0.0	0.0	0.0	0.0	0.0	0.0	0.2940	0.2730	0.2520	0.2310	0.0058	8.393	0.2523
9	0.0	0.0	0.0	0.0	0.0	0.0	0.0	0.0	0.5355	0.5145	0.0051	9.489	0.5003
Peaked													
2	0.0525	0.6125	0.3325	0.0525							0.0207	2.358	0.4481
3	0.0350	0.3710	0.2870	0.2030	0.1190	0.0350					0.0158	3.078	0.2547
4	0.0262	0.1462	0.2663	0.2183	0.1702	0.1222	0.0742	0.0262			0.0119	4.068	0.1742
5	0.0210	0.1143	0.2077	0.1810	0.1543	0.1277	0.1010	0.0743	0.0477	0.0210	0.0101	4.787	0.1355
6	0.0	0.0	0.0262	0.2663	0.2262	0.1863	0.1462	0.1063	0.0662	0.0262	0.0084	5.809	0.1786
7	0.0	0.0	0.0	0.0	0.0350	0.3710	0.2870	0.2030	0.1190	0.0350	0.0069	7.090	0.2532
8	0.0	0.0	0.0	0.0	0.0	0.0	0.0525	0.6125	0.3325	0.0525	0.0058	8.364	0.4463

TABLE 15.3.2. Columns show the average net fertility (ϕ_i) for all ages. Other entries are as in preceding table

tribution can have appreciable effect on stochastic growth rate (Slade and Levenson 1984).

B. *Formulae*

Three patterns of pairwise correlation are worth highlighting, along with the corresponding analytical approximations.

1. *complete independence of ϕ_{it} values.* In this case,

$$a \approx r_0 - \frac{c^2\left(\sum G_i^2\right)}{2T_0^2} \tag{15.3.10}$$

where c denotes the age-invariant coefficient of variation of ϕ_{it} values, G_i is $\phi_i \lambda_0^{-i}$, and T_0 is mean generation length. This case most simply corresponds to variability of the underlying fertility values, *not* to variability of the survivorships since these are, by definition, likely to be positively correlated. In this case one can directly see the dynamical consequences of varying degrees of "diversity" of the ϕ_i values. This follows because $\sum G_i = 1.0$ and therefore, $D = \sum G_i^2$ is a measure of the diversity of the reproductive schedule similar to Simpson's index for species abundances. This index decreases as reproduction is spread over more ages, and for a life history of length ω, has a minimum value of $1/\omega$ and a maximum value of 1. The former corresponds (if λ_0 is 1.0) to a flat ϕ_i schedule with reproduction at all ages, and the latter to reproduction at one age only. Values of D are shown in Tables 15.3.1 and 15.3.2.

2. *a correlation between each pair of ϕ_{it} values of* +1.This is just the case of (15.3.3) in Section 3.1 above.

3. *an age-invariant correlation r between ϕ_{it} values,*

$$a \approx r_0 - \frac{c^2\left(\sum G_i^2 + r \sum\sum G_i G_k\right)}{2T_0^2}, \quad i \neq k. \tag{15.3.11}$$

In this case, r may take on any value between -1 and $+1$. It is important to note that life histories of different lengths must differ in the degree to which their ϕ_{it} values can be negatively correlated. In particular, it is straightforward to show that

$$r \geq \frac{-1}{(n_\phi - 1)} \tag{15.3.12}$$

is a necessary lower bound on the correlation among the ϕ_{it} values. (n_ϕ is the number of varying ϕ_i values in the life history). Thus, assuming age-invariant correlations, the most negative correlation for a life history with two varying ϕ_{it}s is -1.0. In contrast, the most negative correlation possible for any life history with ten varying ϕ_{it}s, for example, is approximately -0.11.

C. *Numerical results*

Figures 15.3.1 to 15.3.6 show the results of polynomial fits to the relationship between the stochastic growth rates and the associated coefficients of variation. Stochastic growth rates for specific cases of correlation among ϕ_{it} values are shown, most of which contain both analytically and numerically derived "indifference" curves. (The lines in these figures are not meant to imply the presence of intermediate life histories (*i.e.*, the abscissa represents a categorical variable). Such curves indicate those life histories which are selectively neutral with respect to one another. It is clear that very distinct life histories can be so classified. For example, when a is equal to 0.002, a life history with early reproduction and short lifetime ($\phi_i = 0.525$, $i = 1, 2$) has no selective advantage over a life history with delayed reproduction in a long lifetime ($\phi_i = 0.0$, $i = 1, \ldots, 8$; $\phi_j = 0.525$, $j = 9, 10$) *if* it differs in the amount of environmental variability it experiences. In particular, the values of c producing neutrality when ϕ_{it}s are completely correlated (Figure 15.3.6, analytical) are approximately 0.37 and 0.75 for the early-producing and late-reproducing life histories, respectively. This result indicates that it is important to determine whether life histories are differentially sensitive to a given amount of environmental variability. Acquisition of such data and information on temporal variability of vital rates (see Baker *et al.* 1981 for a rare example) should be an important goal of experimental evolutionary genetics.

There is no consistent relationship among life histories with respect to the coefficients of variation allowing neutrality. c can increase (Figure 15.3.1, $r = -0.95$, $a = 0.0$), have an intermediate peak (Figure 15.3.3, $a = 0.0$), or decrease (Figure 15.3.5, $a = 0.014$). The consequence is that predictions about the direction of life history evolution must also be based on adequate quantitative information about the quality of the environment (as indexed here by the stochastic growth rate).

Figure 15.3.4 contains analytical indifference curves for the declining and peaked life histories when the ϕ_{it}s are independent. Comparison with Figure 15.3.3 indicates the general similarity of indifference curves despite changes in the geometries of the life histories. Similar results hold for other values of correlation between the ϕ_{it}s.

The analytical data can be plotted in a different manner, as in Figures 15.3.7 – 15.3.9, to reveal how natural selection acts upon life histories with the *same* sensitivity to environmental fluctuations (*i.e.*,a given value of c).

These results identify three dynamical regimes associated with small, intermediate, and high levels of environmental variability as defined by different values of c. We define these terms relative to the correlation structure. Accordingly, for a given c value, a life history with completely correlated ϕ_{it}s experiences more variation than one with independent ϕ_{it}s. Consider, for example, Figure 15.3.8, the case $r = 0$.

1. *Small variability.* When c is relatively low (*e.g.*, $\lesssim 0.5$), there is a

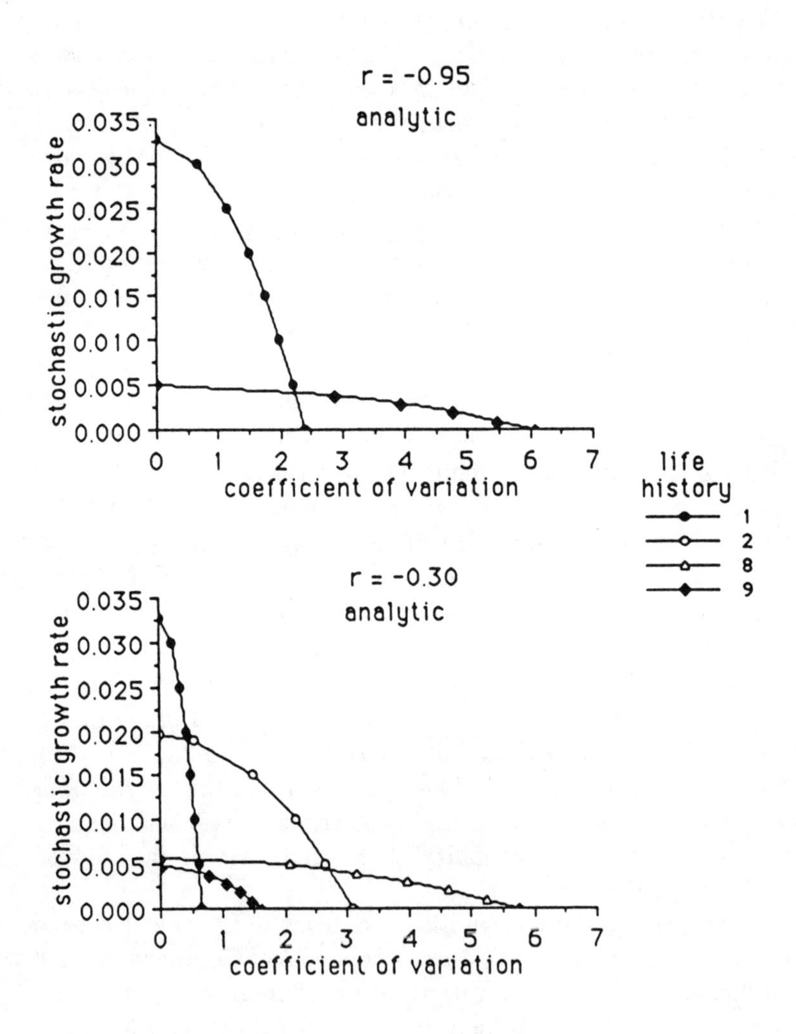

FIGURE 15.3.1. Analytic indifference curves when net fertilities have correlations of -0.95 (*top*) and -0.30 (*bottom*) for flat average life-histories

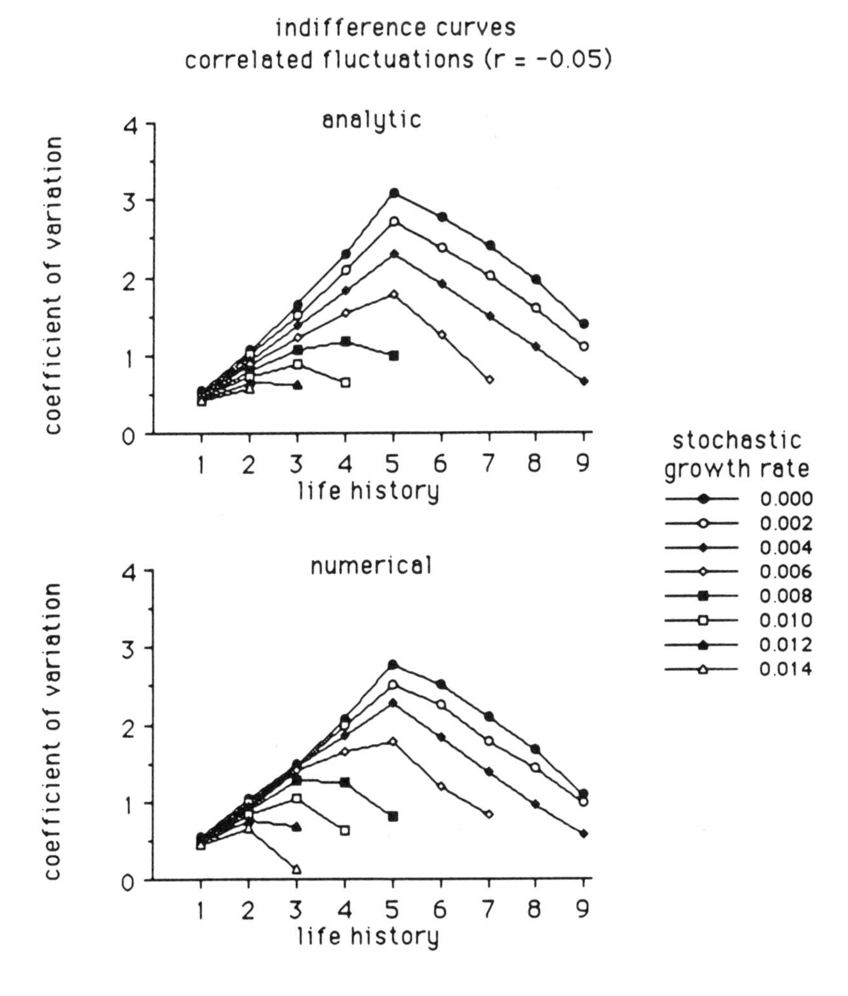

FIGURE 15.3.2. Indifference curves when net fertilities have correlations of -0.05 for flat average life-histories. *Top*, analytical values. *Bottom*, numerical estimates from a polynomial fit to simulation results

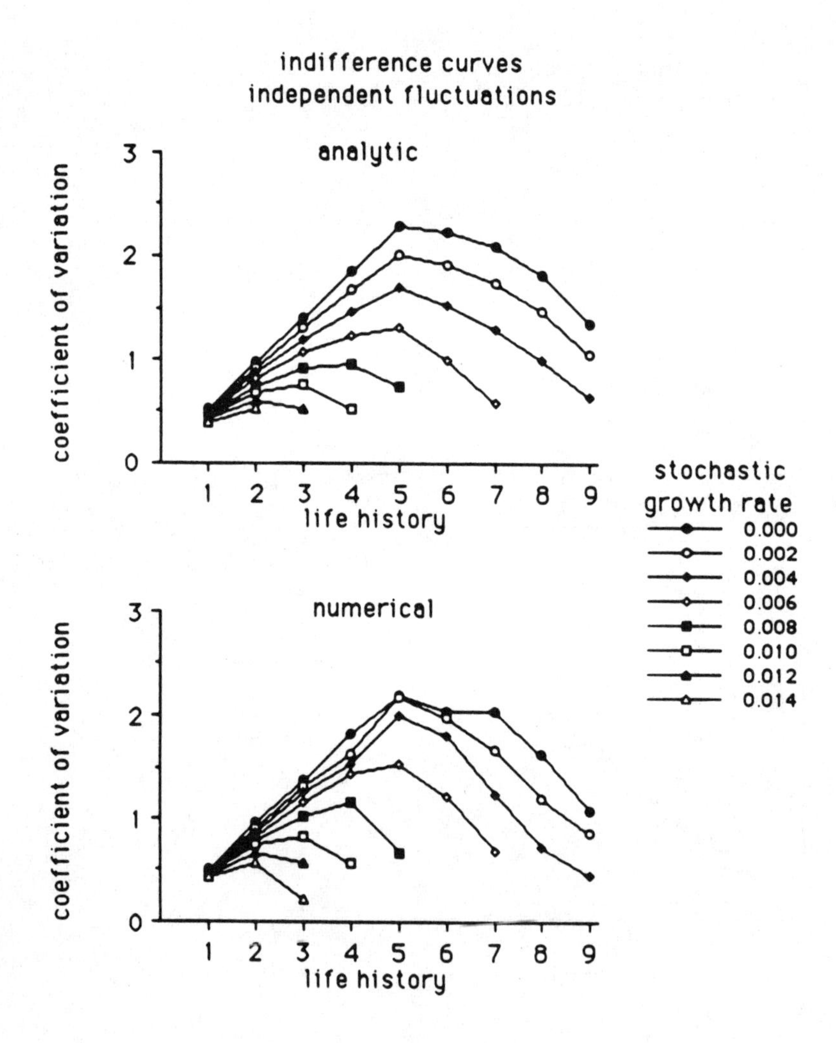

FIGURE 15.3.3. Indifference curves when net fertilities vary independently for flat average life-histories. *Top*, analytical values. *Bottom*, numerical estimates from a polynomial fit to simulation results

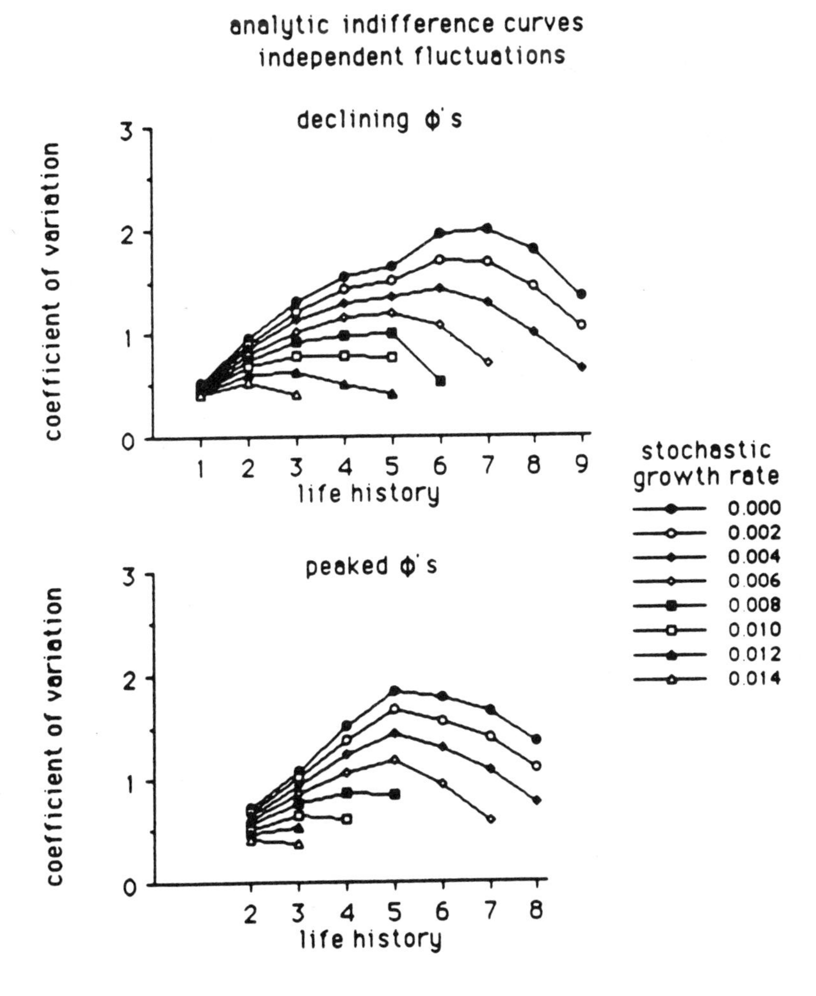

FIGURE 15.3.4. Indifference curves when net fertilities vary independently, calculated analytically. *Top*, declining average life histories. *Bottom*, peaked average life histories

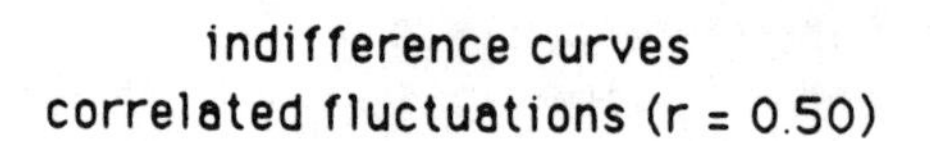

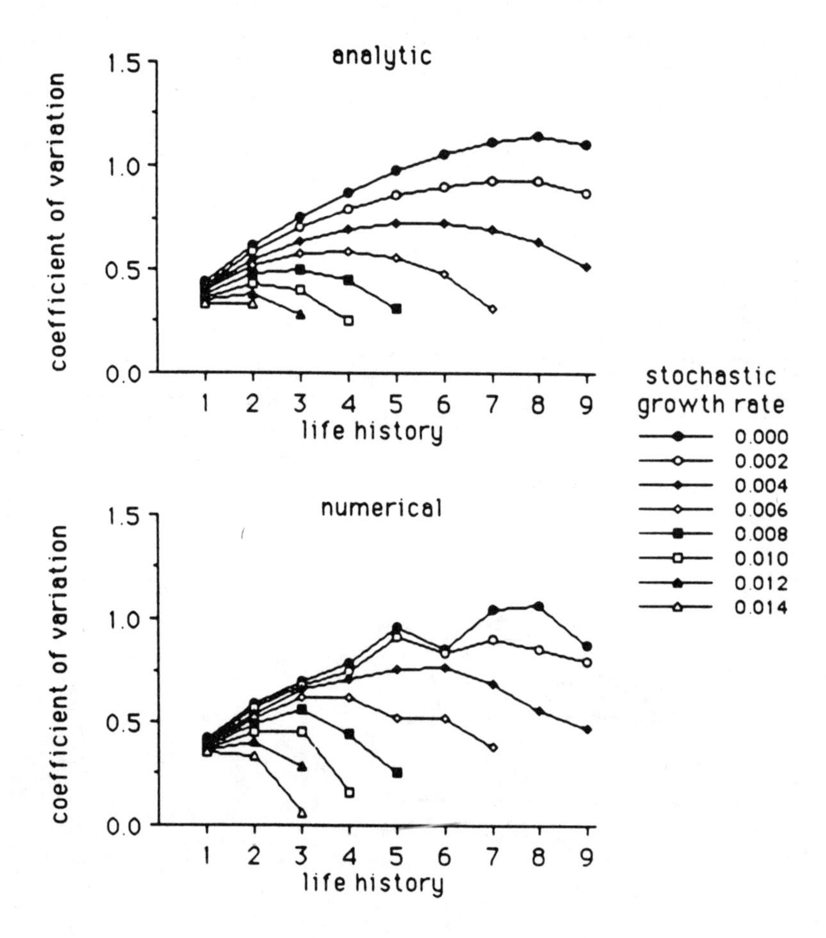

FIGURE 15.3.5. Indifference curves when net fertilities have correlations of 0.50 for flat average life-histories. *Top*, analytical values. *Bottom*, numerical estimates from a polynomial fit to simulation results

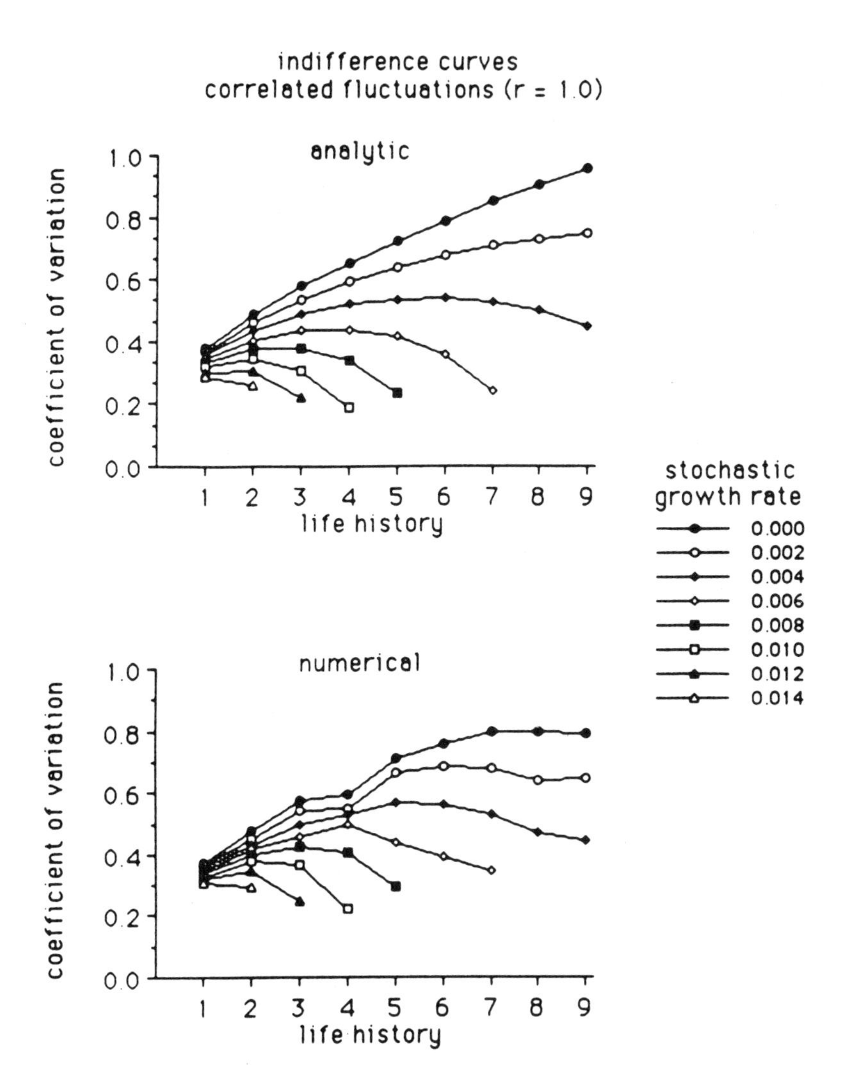

FIGURE 15.3.6. Indifference curves when net fertilities have correlations of 1.0 for flat average life-histories. *Top*, analytical values. *Bottom*, numerical estimates from a polynomial fit to simulation results

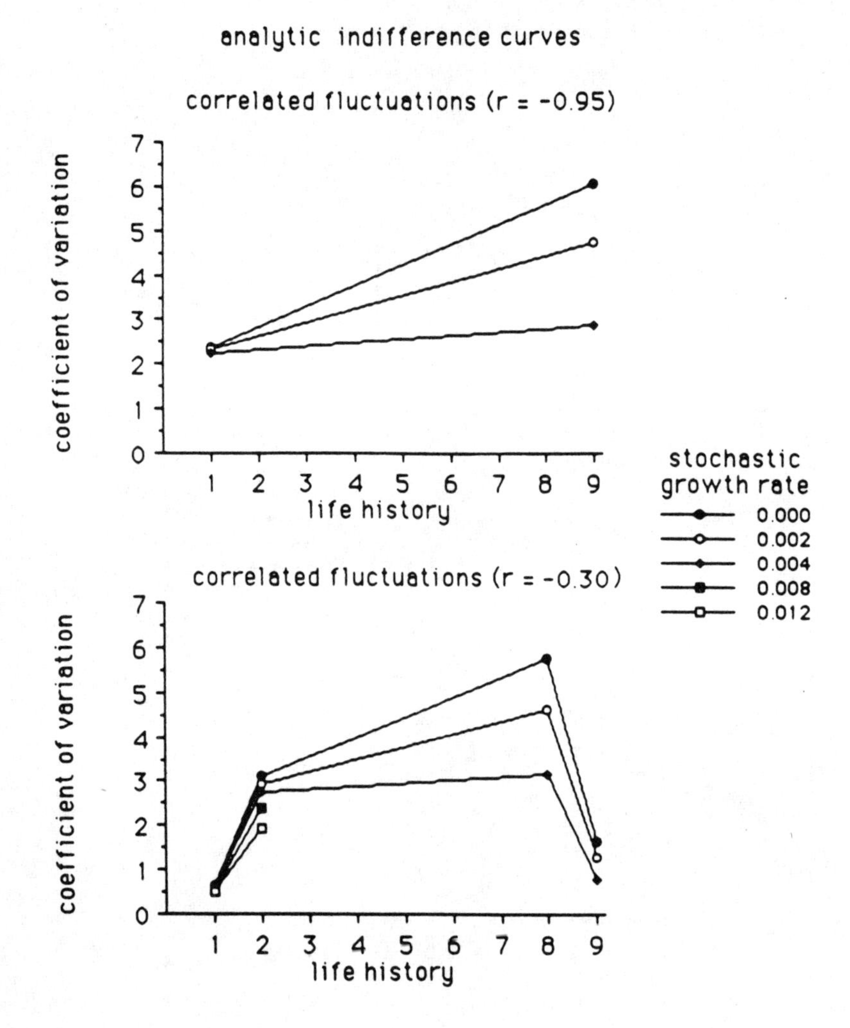

FIGURE 15.3.7. Analytic stochastic growth rate for increasing variability when net fertilities have correlations of -0.95 (*top*) and -0.30 (*bottom*) for flat average life-histories

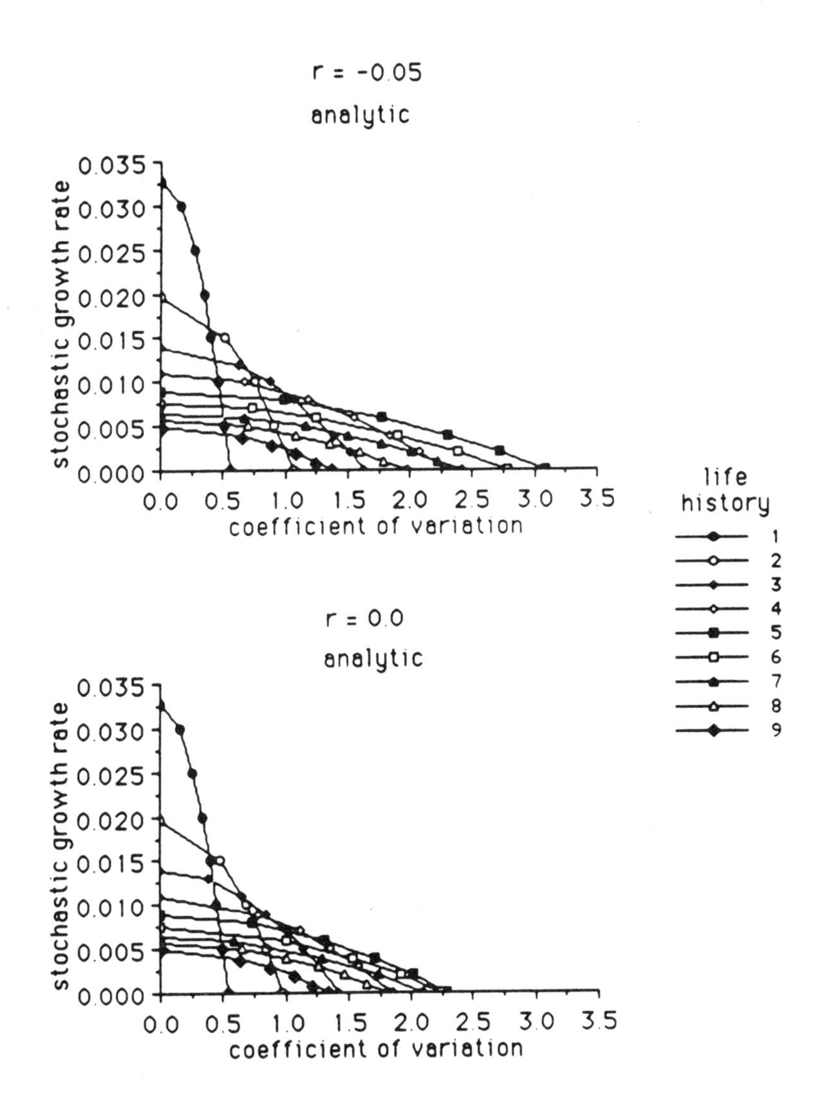

FIGURE 15.3.8. Analytic stochastic growth rate for increasing variability when net fertilities have correlations of -0.05 (*top*) and 0.0 (*bottom*) for flat average life-histories

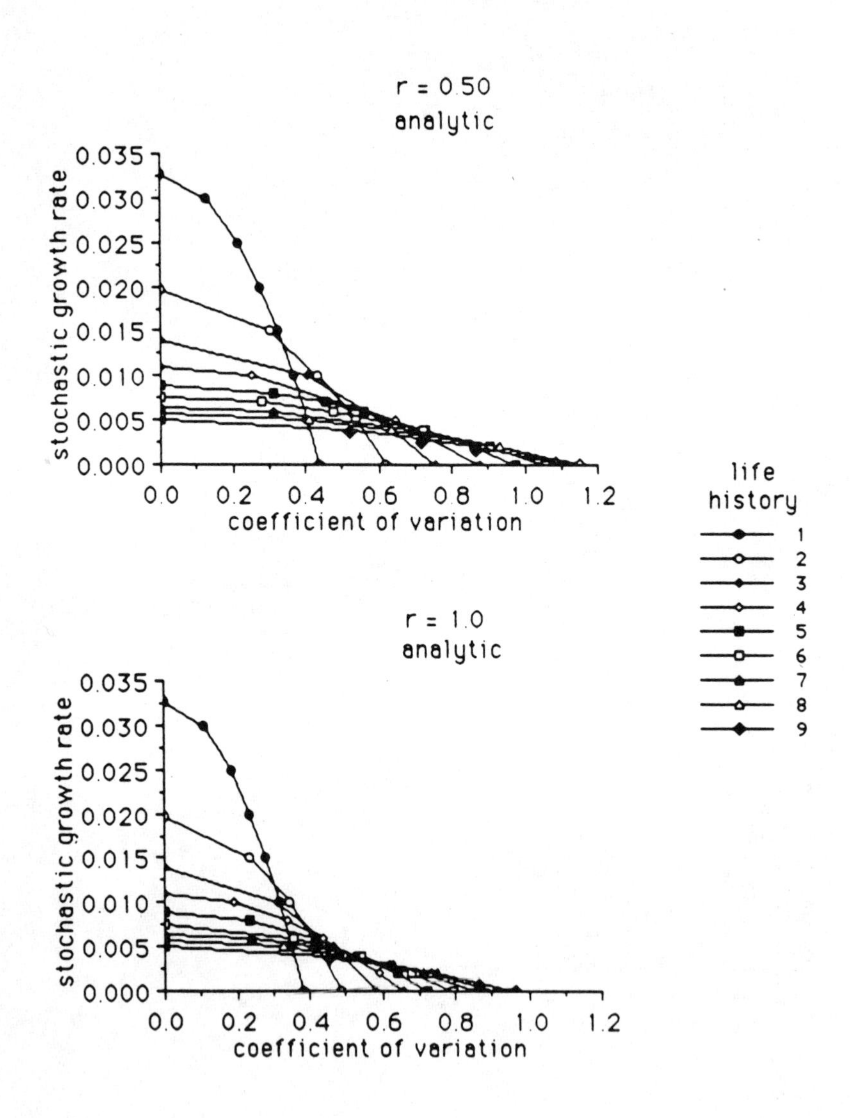

FIGURE 15.3.9. Analytic stochastic growth rate for increasing variability when net fertilities have correlations of 0.50 (*top*) and 1.0 (*bottom*) for flat average life-histories

clear selective advantage to a type of semelparity—reproducing early in a short lifetime. This is basically Cole's result in an age-structured, stochastic context.

2. *Intermediate variability.* Life histories with various types of iteroparous reproductive schedules have an advantage relative to genotypes with larger $\log \lambda_0$ or T_0 values. For example, when $c = 0.75$, life history 3 can invade as a heterozygous mutant into a monomorphic population composed of any other life history. Note that the entire set of life histories has similar stochastic growth rates when $c \geq\approx 1.0$ and $\leq\approx 1.5$, and the outcome of life history evolution will depend closely on the details.

3. *High variability.* For larger c values ($\geq\approx 1.5$), there is an advantage to the most iteroparous life history. Note that when ϕ_{it}s are positively correlated, long-lived, late-reproducing genotypes can also (depending on the value of r) have the highest stochastic growth rates. Consequently, selection may favor dispersed *or* delayed reproduction in highly variable environments.

The absolute amount of environmental variability and the correlation of ϕ_{it}s can be critical to predicting the direction of life history evolution in a particular instance. For example, when $c = 1.0$, the life history in Table 15.3.1 with the highest value of a is 1 when $r = -0.95$, 2 when $r = -0.30$, 3 when $r = -0.05$, 4 when $r = 0.00$, 8 when $r = 0.5$, and 9 when $r = 1.0$. When $c = 0.25$, life history 1 has the highest value of a, *regardless* of the value of r.

3.3 Some general conclusions

One evolutionary implication to be drawn from indifference curves concerns the comparative analysis of life histories within or between conspecific populations or between closely related species. These results clearly reveal the potential for the nonselective (and probably transitory) coexistence of distinct life histories within a population as a result of, say, mutational accumulation. One consequence of this accumulation is that even marked differentiation of life histories between populations or between closely related species may be caused by random fixation of alleles due to stochastic fluctuations in population size or genetic drift. It is not appropriate to assume, *a priori*, that such differentiation reflects adaptation to different environments (*e.g.*, Gadgil and Bossert 1970, Schaal 1984). The same point has been made using a different approach by Schaffer (1974a).

Now recall the condition (equation 6.1.8) for a protected polymorphism. When heterozygous individuals are intermediate in average life history between constituent homozygotes in the crossover regime of intermediate variability, it is clearly possible to have selectively maintained polymorphisms.

That such polymorphisms are likely in variable environments is obviously suggested by the indifference curves, and echoes the findings of many empirical studies (Denno and Dingle 1981).

These results also bear on the relationship between plasticity and genetic polymorphism (Bradshaw 1965, Jain 1979, Scheiner and Goodnight 1984). Phenotypic plasticity is measured by the individual coefficients of variation for vital rates and the pairwise correlation between vital rates. Since the growth rate a depends on sensitively on these measures of plasticity, and in turn a determines conditions for polymorphism, there is a direct relationship implied by the theory. As pointed out by Scheiner and Goodnight, the relationship is not necessarily antagonistic. Selection can be coupled: for example, if plasticity were to increase in a variable environment, there would be an increase in c and in increasing advantage to iteroparous life histories. In such a case, genetic variation might decrease as plasticity increases. However, an increase in plasticity which pushed a population from low c to intermediate c regimes would open the population to invasion by a wide range of life histories, and here increasing plasticity could drive an increase in genetic polymorphism.

Another interesting aspect of life history evolution is that there can be selection between genotypes purely on the basis of the *sign* of the covariance between vital rates. For example, assume that environmental variability affects only fertilities but does so equally for all genotypes. Then, a mutant heterozygote with negative correlations among fertilities can enter a population composed of a genotype with *identical* average fertilities which are independent or positively correlated. This point underscores the multivariate nature of "fitness" in age-structured populations and is also a reminder that negative correlations between life history components can be selectively advantageous. Hence, care must be taken during analysis of "costs" of reproduction (see Reznick (1985) for a general discussion) to separate such correlations from those due to a nonselective constraint on reproduction and survival.

Finally, compare these results with Schaffer's (1974b) important attempt to understand life history evolution in variable environments. He analyzes how dichotomous variation of an age-invariant average litter size (B) or average adult survivorship rate (P) affects the geometric mean of the growth rates ($B + P$). If variability affects B, the optimal life history is more iteroparous). This case includes the consequences of variable juvenile survivorship since B is the "effective litter size" and is a product of adult fertilities and juvenile survivorship (as in a ϕ_{it} value). Conversely, if P is affected by variability the optimal life history has more concentrated early reproduction (*i.e.*, is more semelparous).

Schaffer's results are most usefully viewed as applying to a special kind of life history in which the average effective litter size and age-specific survivorship are age-invariant in a long lifetime (see Tuljapurkar 1982b). However, his results are not general because the geometric mean he uses will not equal the stochastic growth rate for other kinds of life history. What results does the more general analysis produce which are not seen in Schaffer's special analysis?

Variance only in B corresponds to a life history in which all nonzero ϕ_i values vary temporally but decline with age due to the fixed average survivorship (p_i). Hence, the results in Figure 15.3.4 concerning declining life histories apply. So, for example, when $c = 2.0$, life history 7 (see Table 15.3.2) has the highest value of a. When $c = 1.4$, life histories 4–7 have nearly identical a values. It is clear that variance in B does not necessarily select for dispersed reproduction.

Variance only in P corresponds to life histories in which $\phi_{\alpha t}$ is nonzero but does not vary. All other ϕ_{it} vary. Therefore, equations 15.3.10–12, which assume that all ϕ_{it} vary, do not apply. The effect of this difference for a given life history is to increase the coefficient of variation necessary to achieve a particular stochastic growth rate but the qualitative results of the analysis remain unchanged. Hence, variance in P does not necessarily select for concentrated reproduction. Clearly, Schaffer's results do not carry over to more general classes of life history.

4 Iteroparity in Fish: Murphy (1968) Revisited

A nice biological application of the results of Section 15.3 is to the evolution of reproductive characteristics of fish. Murphy (1968) compared the dynamics of egg-laying fish which are subject to unpredictable random variation in juvenile survival. He argued that as the level of variability increased there would be greater advantage to iteroparity. In Section 15.2, I argued for the same point using the theory of random vital rates. The theory has at least two advantages: it shows that the argument does not hinge on a fortuitous choice of examples, and it provides quantitative relationships applicable to a range of life histories.

To illustrate the quantitative usefulness of the present theory, consider a set of life histories constructed according to the "declining" schedule of average vital rates in Table 15.2. Figure 15.4.1 plots indifference curves (curves of equal a) for all life histories in the set. Now for fish of the sort Murphy considered, the relevant life history set consists of life histories 1-5 (see Table 15.2) in which the progression is towards "increasing iteroparity". Figure 15.4.2 plots an indifference curve for this subset in a different way: with reproductive span ($\omega - \alpha$) on the vertical axis and the squared coefficient of variation on the horizontal axis. The line in Figure 15.4.2 is for $a = 0.001$. This line predicts a relationship between the extent of iteroparity and the level of randomness such that life histories which obey it are equally fit. In nature, the theory predicts that species (preferably of similar ecological habit) which occupy habitats of different variability, or (better yet) populations of one species which occupy such habitats, should display a relationship of the type shown in Figure 15.4.2.

In Figure 15.4.3 I have redrawn the relevant empirical data estimated by Murphy. The agreement with theory is striking. Essentially similar results are obtained for different populations of the American shad *Alosa sapidissima*, based on Carscadden and Leggett (1978). Clearly, detailed analysis of such examples in the context of the theory of random rates will be very useful.

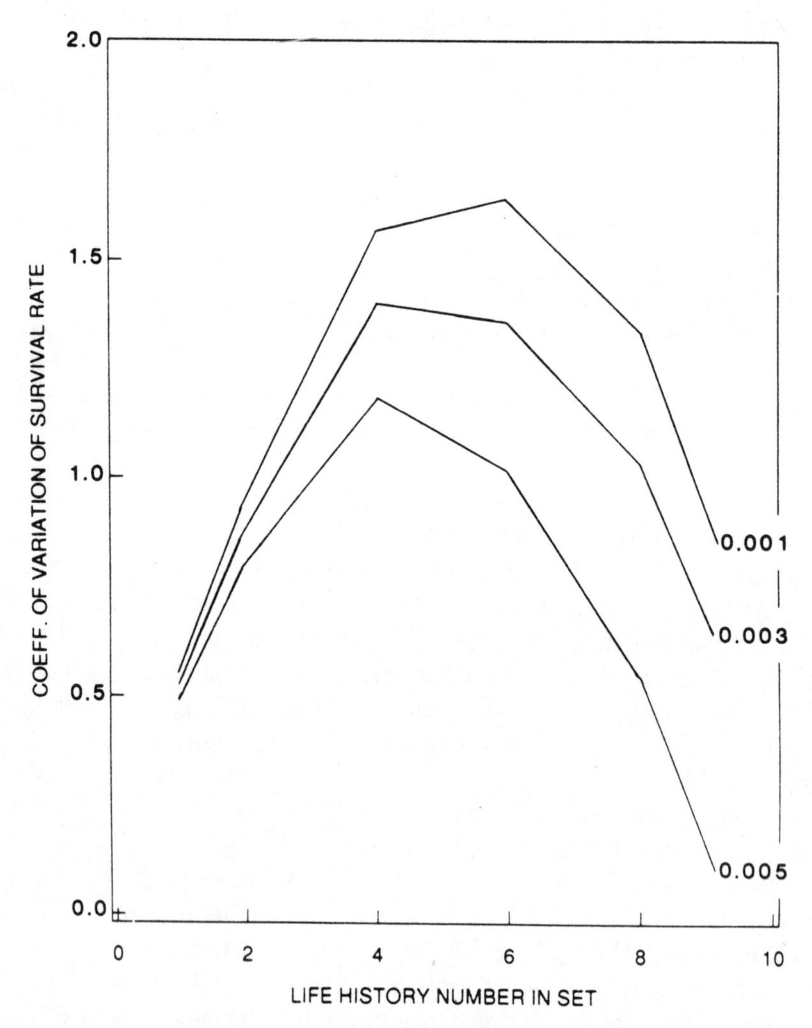

FIGURE 15.4.1. Analytic indifference curves when net fertilities have correlations of 1.0 for declining average life-histories. Curves are labelled with stochastic growth rate

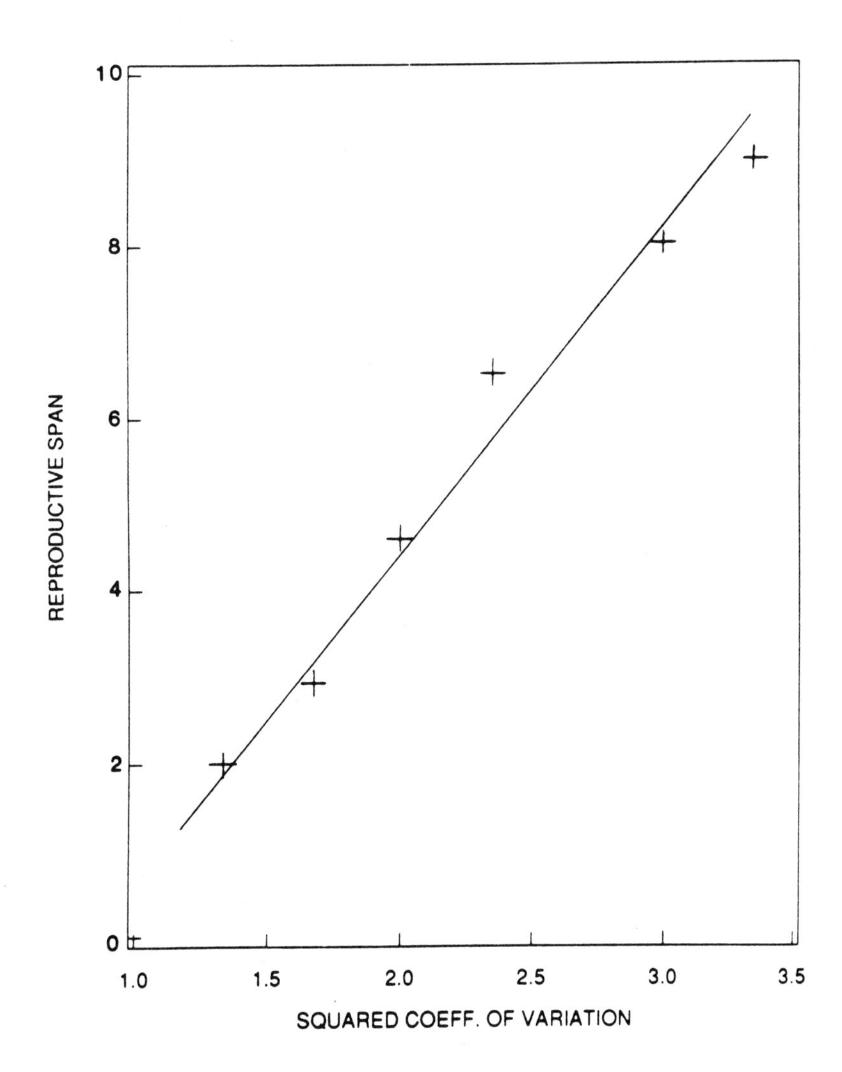

FIGURE 15.4.2. Reproductive span plotted against squared coefficient of variation for a fixed growth rate of 0.001. Straight line is an eyeball fit

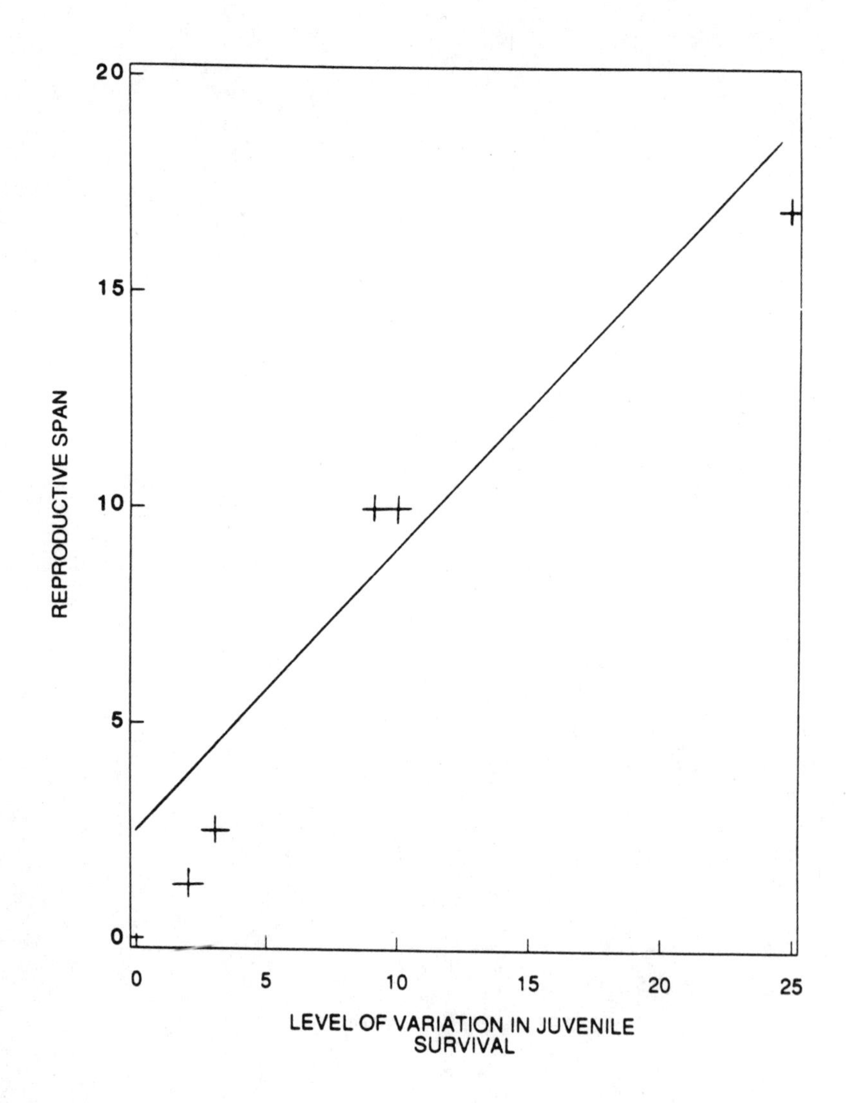

FIGURE 15.4.3. Plot based on Murphy (1968), Figure 2. Reproductive span is measured in years, variability is estimated in survival rate of eggs

16

LIFE HISTORY EVOLUTION: DELAYED FLOWERING

Chapter 5 mentioned the work of Klinkhammer and de Jong on the advantages of delayed flowering in biennials. The species under consideration (such as *Cirsium vulgare*) reproduce by flowering only once and then die. The strict biennial habit involves flowering in the second year of life, but delayed flowering of up to several years is commonly observed. Klinkhammer and de Jong (1983) argued that delayed flowering is akin to iteroparity and evolved as a response to variable reproductive success; they used a computer simulation model. Roerdink (1987, 1989) analyzed their model using the theory of stochastic demography. I discuss below his work and some related questions; this analysis complements the more general discussion of the preceding chapter.

1 Models For Delayed Flowering

I begin by considering a general model for a plant population with no seed pool and with reproduction only in the second or later years of life. Suppose that $N_t(i)$ is the number of plants between i and $i+1$ years old at time t. Let s_i be the fraction of plants which survive from age class i to $i+1$, conditional on their not flowering. Let f be the probability that a plant flowers in any particular year after it enters age class 2. And let F_{it} be the reproductive rate (numbers of offspring plants in age class 1 in year t) per flowering plant in age class i in year $t-1$. Then

$$\begin{aligned} N_{t+1}(1) &= f \sum_{i \geq 2} F_{it+1} N_t(i), \\ N_{t+1}(2) &= s_1 N_t(1) \\ N_{t+1}(i+1) &= (1-f) s_i N_t(i), \quad i \geq 2. \end{aligned}$$

Suppose there are a total of k age classes, and *assume* (i) $s_i = s$ for all i, (ii) $F_{it} = \phi_{t+1}$, all i. Then (16.1.1) can be rewritten using the total number of individuals age 2 or older,

$$W_t = \sum_{i \geq 2} N_t(i), \tag{16.1.1}$$

to get

$$\begin{aligned} N_{t+1}(1) &= f\phi_{t+1}W_t, \\ W_{t+1} &= sN_{t+1}(1) + (1-f)sW_t - (1-f)sN_t(k). \end{aligned}$$

Now *assume* also (iii) $0 < s < 1$ and k large ($k \gg 1$). Then we can argue that $[N_t(k)/W_t]$ is of the order of $s^k \ll 1$ and replace (16.1.3) by the Klinkhammer-de Jong model

$$\begin{aligned} N_{t+1}(1) &= f\phi_{t+1}W_t, \\ W_{t+1} &= sN_{t+1}(1) + (1-f)sW_t. \end{aligned}$$

This model is the focus of the rest of the chapter.

2 Why Delayed Flowering is a Surprise

Suppose now that (16.1.4) is a satisfactory population model and rewrite it in matrix form by setting

$$\boldsymbol{N}_t = \begin{pmatrix} N_t(1) \\ W_t \end{pmatrix}, \tag{16.2.1}$$

$$\boldsymbol{X}_t = \begin{pmatrix} 0 & f\phi_t \\ s & (1-f)s \end{pmatrix}, \tag{16.2.2}$$

to get

$$\boldsymbol{N}_{t+1} = \boldsymbol{X}_{t+1}\boldsymbol{N}_t. \tag{16.2.3}$$

(A notational point: in keeping with the rest of this book the vital rate matrix in (16.2.3) has subscript $t+1$. Roerdink (1987) used t. The results are the same). *Assume* that the ϕ_t are I.I.D. random variables.

Focus first on the *average* population. The average vector $\mathsf{E}(\boldsymbol{N}_t)$ grow according to the average matrix

$$\boldsymbol{b} = \mathsf{E}(\boldsymbol{X}_t) = \begin{pmatrix} 0 & f\langle\phi\rangle \\ s & (1-f)s \end{pmatrix}, \qquad \langle\phi\rangle = \mathsf{E}\,\phi_t \tag{16.2.4}$$

We have $0 \le f \le 1$ and assume $0 < \langle\phi\rangle < \infty$. Notice that the matrix $\boldsymbol{b}$ in (16.2.4) is now nonnegative, is primitive for $0 < f < 1$, and its transpose looks just like a Leslie matrix. The dominant eigenvalue of $\boldsymbol{b}$ is

$$\lambda_0(f) = \frac{1}{2}s(1-f) + \frac{1}{2}[s^2(1-f)^2 + 4f\langle\phi\rangle s]^{\frac{1}{2}}, \tag{16.2.5}$$

and it is easy to check that

$$\frac{\partial\lambda_0}{\partial f} \lesseqqgtr 0 \text{ according as } \langle\phi\rangle \lesseqqgtr s. \tag{16.2.6}$$

Now if $\langle\phi\rangle < s$, $\lambda_0(f) \leq \lambda_0(1) = \sqrt{\langle\phi\rangle s} < 1$. So for a growing population we must have $\langle\phi\rangle > s$ and $\lambda_0(f)$ is maximized when $f = 1$, *i.e.*, in the deterministic theory *flowering should never be delayed in a deterministic world.*

Thus the question: is delayed flowering likely to be a response to randomly varying reproductive success?

3 Analysis of the Model

For the model (16.2.1.–2) we proceed as in Chapter 8 by introducing the ratio

$$U_t = N_t(1)/W_t, \tag{16.3.1}$$

which then satisfies the equation

$$U_{t+1} = \frac{f\phi_{t+1}}{(1-f)s + sU_t}.$$

If we scale U_t and let

$$R_t = U_t/(1-f), \tag{16.3.2}$$

$$z = (1-f)^2 s/f, \tag{16.3.3}$$

then

$$R_{t+1} = \frac{\phi_{t+1}}{z(1+R_t)}. \tag{16.3.4}$$

Observe that (16.3.4) would be identical to (8.2.12) if we were to set ϕ_t equal to $(1/f_t)$. Hence, the analysis of Section 8.2 can be readily adapted to the present model. Roerdink (1987) does precisely this, with the slight generalization that he takes ϕ_t to be I.I.D. random with a two-parameter probability density

$$g(w) = \{k^b/\Gamma(b)\}w^{b-1}e^{kw}, \quad b > 0, k > 0. \tag{16.3.5}$$

This yields

$$\mathsf{E}(\phi_t) = \frac{b}{k} = \langle\phi\rangle, \tag{16.3.6}$$

$$\mathrm{Var}(\phi_t) = \frac{b}{k^2}, \tag{16.3.7}$$

so that $\langle\phi\rangle$ and $\mathrm{Var}(\phi)$ can be varied independently. The calculation of a proceeds exactly as described in Section 8.2 and will not be repeated here.

A separate calculation (Roerdink 1989) provides a numerical expression for the variance σ^2 of the logarithm of population size (*cf.* Section 4.2.3). The latter calculation uses a novel and interesting technique, but is too involved to reproduce here.

4 Biological Results

The actual results require numerical evaluation of integral formulae analogous to (8.2.16). The advantage of having these formulae is that one does not have to do simulations.

The *first* important result is based on computing a for the matrices of (16.2.2) as a function of the flowering fraction f. Recall that the deterministic $\lambda_0(f)$ is maximum at $f = 1$ (when $\lambda_0 \geq 1$). When we compute $a(f)$ the answer depends also on the coefficient of variation

$$c = \{\mathrm{Var}(\phi)/,\langle\phi\rangle^2\}^{\frac{1}{2}} = 1/\sqrt{b}. \tag{16.4.1}$$

For all values of $b < \infty$ the function $a(f)$ is found to have a *maximum* at some $f < 1$. When c *is small* (*e.g.*, $c \leq 0.5$) the value of f for maximal a is *very close* to 1, and thus *biologically indistinguishable* from 1. When c *is large* (*e.g.*, $c \geq 1.5$) the values of f for maximal a is substantially less than 1, and *delayed flowering would successfully "invade"* a population of strict biennials.

The *second* important result is based on knowing the variance σ^2 which, together with a, yields an estimate of the probability of extinction (see Tuljapurkar and Orzack 1980 for the theory behind the estimate). Roerdink finds that this estimated probability of extinction decreases as f increases from 0, reaches a minimum at a value of f between 0 and 1 and then rises as f increases to 1. The minimum is sharper for high values of c.

These results provide considerable justification for the argument that increasing variability in reproductive success leads to an advantage for delayed flowering.

Bibliography

Abramov, L. M. and Rohlin, V. A. (1966). Entropy of a skew-product transformation with invariant measure. *American Mathematical Society Translation Series*, 2:255–265.

Abramowitz, M. and Stegum, I. (1965). *Handbook of mathematical functions.* Dover, New York.

Alho, J. M. and Spencer, B. D. (1985). Uncertain population forecasting. *Journal of the American Statistical Association*, 80:306–314.

Athreya, K. B. and Karlin, S. (1971). On branching processes with random environments. I. Extinction probabilities. *Annals of Mathematical Statistics*, 42:1499–1520.

Baker, M. C., Mewaldt, L. R., and Stewart, R. M. (1981). Demography of white-crowned sparrows (*Zonotrichia leucophrys* Nuttalli). *Ecology*, 62:636–644.

Bartlett, M. S. (1978). *An Introduction to Stochastic Processes.* Cambridge University Press, London.

Begon, M., Harper, J. L., and Townsend, C. R. (1986). *Ecology.* Sinauer Associates, Sunderland, Mass.

Benettin, G., Galgani, L., Giorgilli, A., and Strelcyn, J. M. (1980). Lyapunov characteristic exponents for smooth dynamical systems and for hamiltonian systems: A method for computing all of them. *Meccanica*, 15:9–64.

Benettin, G., Galgani, L., and Strelcyn, J. M. (1976). Kolmogorov entropy and numerical experiments. *Physical Review*, A6:2338–2345.

Bernardelli, H. (1941). Population waves. *Journal of the Burma Research Society*, 31:1–18.

Bharucha, B. H. (1961). On the stability of randomly varying systems. Master's thesis, University of California, Berkeley.

Bierzychudek, P. (1982). The demography of jack-in-the-pulpit, a forest perennial that changes sex. *Ecological Monographs*, 52:335–351.

Billingsley, P. (1968). *Convergence of probability measures.* Wiley, New York.

Box, G. E. P. and Jenkins, G. M. (1970). *Time Series, Forecasting and Control.* Holden Day, San Francisco.

Boyce, M. S. (1977). Population growth with stochastic fluctuations in the life table. *Theoretical Population Biology*, 12:366–373.

Boyce, M. S. (1979). Population projections with fluctuating fertility and survivorship schedules. In *Proceedings of the Summer Computer Simulation Conference*, pages 385–388, Toronto.

Bradshaw, A. D. (1965). Evolutionary significance of phenotypic plasticity in plants. *Advances in Genetics*, 13:115–155.

Bulmer, M. G. (1985). Selection for iteroparity in a variable environment. *The American Naturalist*, 126:63–71.

Caswell, H. (1978). A general formula for the sensitivity of population growth rate to changes in life history parameters. *Theoretical Population Biology*, 14:215–230.

Caswell, H. (1989). *Matrix population models: construction, analysis, and interpretation.* Sinauer Associates, Sunderland, Mass.

Caughley, G. (1977). *Analysis of vertebrate populations.* Wiley, New York.

Charlesworth, B. (1980). *Evolution in age-structured populations.* Cambridge University Press.

Coale, A. J. (1957). How the age distribution of a human population is determined. In *Cold Spring Harbor Symposium on Quantitative Biology*, volume 22, pages 83–87.

Coale, A. J. (1972). *The growth and structure of human populations: A mathematical investigation.* Princeton University Press.

Cohen, J. E. (1976). Ergodicity of age structure in populations with Markovian vital rates, I:Countable states. *Journal of the American Statistical Association*, 71:335–339.

Cohen, J. E. (1977a). Ergodicity of age structure in populations with Markovian vital rates. II. general states. *Advances in Applied Probability*, 9:18–37.

Cohen, J. E. (1977b). Ergodicity of age structure in populations with Markovian vital rates, iii: Finite-state moments and growth rate; an illustration. *Advances in Applied Probability*, 9:462–475.

Cohen, J. E. (1979a). Comparative statics and stochastic dynamics of age-structured populations. *Theoretical Population Biology*, 16:159–171.

Cohen, J. E. (1979b). Contractive inhomogeneous products of non-negative matrices. *Mathematical Proceedings of the Cambridge Philosophical Society*, 86:351–364.

Cohen, J. E. (1979c). Ergodic theorems of demography. *Bulletin of the American Mathematical Society N.S.*, 1:275–295.

Cohen, J. E. (1979d). Long-run growth rates of discrete multiplicative processes in Markovian environments. *Journal of mathematical analysis and applications*, 69:243–251.

Cohen, J. E. (1980). Convexity properties of products of random non-negative matrices. *Proceedings of the National Academy of Sciences*, 77:3749–3752.

Cohen, J. E. (1982). Multi regional age-structured populations with changing vital rates: weak and stochastic ergodic theorems. In Land, K. C. and Rogers, A., editors, *Multi regional mathematical demography*, pages 477–503. Academic Press, New York.

Cohen, J. E. (1986). Population forecasts and confidence intervals for Sweden: A comparison of model-based and empirical approaches. *Demography*, 23:105–126.

Cohen, J. E., Christensen, S. W., and Goodyear, C. P. (1983). An age-structured fish population model with random survival of eggs: calculation of asymptotic growth rates and application to Potomac River striped bass. *Canadian Journal of Fisheries and Aquatic Sciences*, 40:2170–2183.

Cohen, J. E., Kesten, H., and Newman, C. M., editors (1986). *Random matrices and their applications*, volume 50. Contemporary Mathematics.

Cole, L. C. (1954). The population consequences of life history phenomena. *Quarterly Review of Biology*, 29:103–137.

Cull, P. and Vogt, A. (1973). Mathematical analysis of the asymptotic behavior of theLeslie matrix population model. *Bulletin of Mathematical Biology*, 35:645–661.

de Jong, T. J. and Klinkhammer, P. G. L. (1988a). Population ecology of the biennials *Cirsium vulgare* and *Cynoglossum officianale* in a coastal sand-dune area. *Journal of Ecology*, 76:366–382.

de Jong, T. J. and Klinkhammer, P. G. L. (1988b). Seedling establishment of the biennials *Cirsium vulgare* and *Cynoglossum officianale* in a sand-dune area: the importance of wtaer for differential survival and growth. *Journal of Ecology*, 76:383–392.

Demetrius, L. (1987). Random spin models and chemical kinetics. *Journal of Chemical Physics*, 87:6939–6946.

den Boer, P. J. (1981). On the survival of a population in an heterogeneous and variable environment. *Oecologia*, 50:39–53.

Denno, R. F. and Dingle, H. (1981). *Insect Life History Patterns*. Springer-Verlag, New York.

Dyson, F. J. (1953). The dynamics of a disordered linear chain. *The Physical Review*, 92:1331–1338.

Ellner, S. (1986). The accuracy of Bartlett's small-fluctuation approximation for stochastic-difference-equation population models. *Mathematical Biosciences*, 74:233–246.

Fisher, R. A. (1930). *The Genetical Theory of Natural Selection*. Dover, New York, 2 edition.

Furstenberg, H. and Kesten, H. (1960). Products of random matrices. *Annals of Mathematical Statistics*, 31:457–469.

Gadgil, M. (1971). Dispersal: population consequences and evolution. *Ecology*, 52:253–260.

Gadgil, M. and Bossert, W. H. (1970). Life history consequences of natural selection. *The American Naturalist*, 104:1–24.

Gerrodette, T., Goodman, D., and Barlow, J. (1985). Confidence limits for population projections when vital rates vary randomly. *Fishery Bulletin*, 83:207–217. (Washington).

Giesel, J. T. (1976). Reproductive strategies as adaptations to life in temporally heterogeneous environments. *Annual Reviews of Ecology and Systematics*, 7:57–80.

Gillespie, J. H. (1977). Natural selection for variances in offspring numbers: a new evolutionary principle. *The American Naturalist*, 111:1010–1014.

Golubitsky, M., Keeler, E. B., and Rothschild, M. (1976). Convergence of the age-structure: Application of the projective metric. *Theoretical Population Biology*, 7:84–93.

Goodman, D. (1984). Risk spreading as an adaptive strategy in iteroparous life histories. *Theoretical Population Biology*, 25:1–20.

Goodman, L. A. (1969). The analysis of population growth when birth and death depend upon several factors. *Biometrics*, 25:659–681.

Goodyear, C. P., Cohen, J. E., and Christensen, S. W. (1985). Maryland striped bass: Recruitment declining below replacement. *Transactions of the American Fisheries Society*, 114:146–151.

Grime, J. P. (1979). *Plant strategies and negative processes.* John Wiley & Sons, Chichester.

Groenandael, J. M. V. and Slim, P. (1988). The contrasting dyanmics of two populations of *Plantago lanceolata* classified by age and size. *Journal of Ecology*, 76:585–599.

Hajnal, J. (1976). On products of non-negative matrices. *Mathematical Proceedings of the Cambridge Philosophical Society*, 79:521–530.

Haldane, J. B. S. and Jayakar, S. D. (1963). Polymorphism due to selection of varying direction. *Journal of Genetics*, 58:237–242.

Hansen, P. E. (1983). Raising Leslie matrices to powers: a method and application to demography. *Journal of Mathematical Biology*, 18:149–161.

Harper, J. L. (1977). *Population Biology of Plants.* Academic Press, London, New York.

Hasminskii, R. Z. (1980). *Stochastic stability of differential equations.* Sijthoff and Noordhoff, Alphen aan den Rijn. Translation of the Russian edition, Nauka, Moscow, 1969.

Hastings, A. and Caswell, H. (1979). Role of environmental variability in the evolution of life history strategies. *Proceedings of the National Academy of Sciences*, 76:4700–4703.

Heyde, C. C. (1985). On inference for demographic projection of small populations. In LeCam, L. M. and Olshen, R., editors, *Proceedings of the Berkeley Conference in honor of Jerzy Neyman and J. Kiefer*, pages 215–223. Wadsworth and Hayward, Institute of Mathematical Statistics, Monterey.

Heyde, C. C. and Cohen, J. E. (1985). Confidence intervals for demographic projections based on products of random matrices. *Theoretical Population Biology*, 27:120–153.

Holgate, P. (1967). Population survival and life history phenomena. *Theoretical Population Biology*, 14:11–10.

Huenneke, L. F. (1987). Demography of a clonal shrub *Alnus incana ssp. rugosa (Betulaceae). American Midland Naturalist*, 117:43–55.

Huenneke, L. F. and Marks, P. L. (1987). Stem dynamics of a clonal shrub: size transition matrix models. *Ecology*, 68:1234–1242.

Istock, C. A. (1967). The evolution of complex life cycle phenomena: an ecological perspective. *Evolution*, 21:592–605.

Istock, C. A. (1981). Natural selection and life history variation: theory plus lessons from a mosquito. In (Denno and Dingle, 1981), pages 113–127.

Ito, Y. (1980). *Comparative ecology.* Cambridge University Press, Cambridge.

Jackson, J. B. C., Buss, L. W., and Cook, R. E. (1985). *Population biology and evolution of clonal organisms.* Yale University Press, New Haven.

Jain, S. (1979). Adaptive strategies: polymorphism, plasticity and homeostasis. In Solbrig, O. T., Jain, S., Johnson, G. B., and Raven, P., editors, *Topics in Plant Population Biology*, pages 160–187. Columbia University Press, New York.

Johnson, N. L. and Kotz, S. L. (1970). *Continuous univariate distributions*, volume 1. Wiley, New York.

Karlin, S. (1982). Classifications of selection-migration structures and conditions for a protected polymorphism. In Hecht, M. K., Wallace, B., and Prance, G. T., editors, *Evolutionary Biology*, volume 14, pages 61–204. Plenum.

Karlin, S. and Lieberman, U. (1974). Random temporal variation in selection intensities. I. Case of large population size. *Theoretical Population Biology*, 6:355–382.

Karlin, S. and Taylor, H. M. (1975). *A first course in stochastic processes.* Academic Press, New York.

Kato, T. (1966). *Perturbation theory of linear operations.* Springer-Verlag, New York.

Key, E. S. (1987). Computable examples of the maximal Lyapunov exponent. *Probability theory and related fields*, 75:97–107.

Keyfitz, N. (1968). *Introduction to the Mathematics of Population.* Addison Wesley, Reading, Mass.

Keyfitz, N. (1986). The pension question and the problem of demographic uncertainty. Technical report, Food Research Institute, Stanford University, Stanford, CA.

Keyfitz, N. and Flieger, W. (1968). *World Population.* University of Chicago Press, Chicago.

Kim, Y. J. (1987). Dynamics of populations with changing vital rates: generalization of the stable population theory. *Theoretical Population Biology*, 31:306–322.

Kim, Y. J. and Sykes, Z. M. (1976). An experimental study of weak ergodicity in human populations. *Theoretical Population Biology*, 10:150–172.

Kim, Y. J. and Sykes, Z. M. (1978). Dynamics of some special populations with $NRR = 1$. *Demography*, 15:559–569.

Klinkhammer, P. G. L. and de Jong, T. J. (1983). Is it profitable for biennials to live longer than two years? *Ecological Modelling*, 20:223–232.

Kushner, H. J. (1966). Stability of stochastic dynamical systems. *Advances in control systems*, 4:73–102.

Lacey, E. P., Real, L., Antonovics, J., and Heckel, D. G. (1983). Variance models in the study of life histories. *The American Naturalist*, 122:114–131.

Lancaster, P. (1969). *Theory of Matrices.* Academic, New York.

Lande, R. (1982). A quantitative genetic theory of life history evolution. *Ecology*, 63:607–615.

Lande, R. (1987). Extinction thresholds in demographic models of terrestrial populations. *The American Naturalist*, 130:624–635.

Lange, K. (1979). On Cohen's stochastic generalization of the strong ergodic theorem of demography. *Journal of Applied Probability*, 16:496–504.

Lange, K. and Hargrove, J. (1980). Stochastic stable population growth. *Journal of Applied Probability*, 52:289–301.

Law, R. A. (1983). A model for the dynamics of a plant population containing individuals classified by age and size. *Ecology*, 64:224–230.

Le Bras, H. (1971). Eléments pour une théorie des populations instables. *Population*, 26:525–572.

Lee, R. D. (1974). Forecasting births in post-transition populations. *Journal of the American Statistical Association*, 69:607–614.

Lee, R. D. (1977). Forecasting fertility and female labor force participation: An assessment of prospects. Report to U. S. Bureau of Labor Statistics.

Leggett, W. C. and Carscadden, J. E. (1978). Latitudinal variation in reproductive characteristics of american shad (*Alosa sapidissima*):evidence for population specific life history strategies in fish. *Journal of the Fisheries Research Board of Canada*, 35:1469–1478.

Leslie, P. H. (1945). On the use of matrices in certain population mathematics. *Biometrika*, 33:213–245.

Lewontin, R. C. (1965). Selection for colonizing ability. In Baker, H. G. and Stebbins, G. L., editors, *The Genetics of Colonizing Species*, pages 77–94. Academic Press, New York.

Lewontin, R. C. and Cohen, D. (1969). On population growth in a randomly varying environment. *Proceedings of the National Academy of Sciences*, 62:1056–1060.

Lopez, A. (1961). *Problems in stable population theory.* Office of Population Research. Princeton University Press, Princeton, N. J.

MacArthur, R. H. (1968). Selection for life tables in periodic environments. *The American Naturalist*, 102:381–383.

MacDonald, J. (1979). A time series approach to forecasting Australian live births. *Demography*, 16:575–583.

Mandelbrot, B. (1982). *The fractal geometry of nature.* Freeman, San Francisco.

Millar, J. S. and Zammuto, R. M. (1983). Life histories of mammals: an analysis of life tables. *Ecology*, 64:631–635.

Mode, C. J. and Jacobson, M. E. (1987a). On estimating critical population size for an endangered species in the presence of environmental stochasticity. *Mathematical Biosciences*, 85:185–209.

Mode, C. J. and Jacobson, M. E. (1987b). A study of the impact of environmental stochasticity on extinction probabilities by Monte Carlo integration. *Mathematical Biosciences*, 83:105–125.

Moloney, K. A. (1988). Fine-scale spatial and temporal variation in the demography of a prennial bunchgrass. *Ecology*, 69:1588–1598.

Murphy, G. I. (1968). Pattern in life history and the environment. *The American Naturalist*, 102:391–403.

Napiorkowski, M. and Zaus, U. (1986). Average trajectories and fluctuations from noisy, nonlinear maps. *Journal of Statistical Physics*, 43:349–368.

Norton, H. T. J. (1928). Natural selection and Mendelian variation. *Proceedings of the London Mathematical Society*, 28:1–45.

Orzack, S. H. (1985). Population dynamics in variable environments. V. the genetics of homeostasis revisited. *The American Naturalist*, 125:550–572.

Orzack, S. H. and Tuljapurkar, S. (1989). Population dynamics in variable environments. VII. the demography and evolution of iteroparity. *The American Naturalist*, 133:901–923.

Oseledec, V. I. (1968). A multiplicative ergodic theorem: Lyapunov characteristic numbers for dynamical systems. *Transactions of the Moscow Mathematical Society*, 19:197–231.

Pinero, D. and Sarukhan, J. (1982). Reproductive behavior and its individual variability in a tropical palm, *Astrocaryum mexicanum*. *Journal of Ecology*, 70:461–472.

Pollard, J. H. (1968). A note on multitype Galton-Watson processes with random branching probabilities. *Biometrika*, 55:589–590.

Pollard, J. H. (1973). *Mathematical models for the growth of human populations.* Cambridge University Press.

Raghunathan, M. S. (1979). A proof of Oseledec's multiplicative ergodic theorem. *Israel Journal of Mathematics*, 32:356–367.

Reznick, D. (1985). Cost of reproduction: an evaluation of the empirical evidence. *Oikos*, 44:257–267.

Ripley, B. D. (1987). *Stochastic Simulation.* John Wiley & Sons, New York.

Roerdink, J. B. T. M. (1987). The biennial life strategy in a random environment. *Journal of Mathematical Biology*, 26:199–215.

Roerdink, J. B. T. M. (1989). The biennial life strategy in a random environment. Supplement. *Journal of Mathematical Biology*, 27:309–320.

Roff, D. A. (1075). Population stability and the evolution of dispersal in a heterogeneous environment. *Oecologia*, 19:217–237.

Rose, M. R. and Charlesworth, B. (1981). Genetics of life history in *Drosophila melanogaster*. I. Sib analysis of adult females. *Genetics*, 97:173–186.

Ruelle, D. (1978). *Thermodynamic Formalism.* Addison Wesley, Reading, Mass.

Ruelle, D. (1979). Analyticity properties of the characteristic exponents of random matrix products. *Advances in Mathematics*, 32:68–80.

Saboia, J. M. (1977). ARIMA models for birth forecasting. *Journal of the American Statistical Association*, 72:264–270.

Schaal, B. (1984). Life-history evolution, natural selection, and maternal effects in plant populations. In Dirzo, R. and Sarukhan, J., editors, *Perspectives in plant population ecology*, pages 188–206. Sinauer, Sunderland, Mass.

Schaffer, W. M. (1974a). Optimal reproductive effort in fluctuating environments. *The American Naturalist*, 108:783–790.

Schaffer, W. M. (1974b). Selection for optimal life histories: the effects of age structure. *Ecology*, 55:291–303.

Scheiner, S. M. and Goodnight, C. J. (1984). The comparison of phenotypic plasticity and genetic variation in populations of the grass *Danthonia spicata*. *Evolution*, 38:845–855.

Schmidt, H. (1957). Disordered one-dimensional crystals. *The Physical Review*, 105:425–441.

Seneta, E. (1981). *Non-negative matrices and Markov chains*. Springer-Verlag, New York.

Seneta, E. (1984). On the limiting set of nonnegative matrix products. *Statistics and Probability Letters*, 2:159–163.

Shine, R. and Bull, J. J. (1979). The evolution of live-bearing in snakes and lizards. *The American Naturalist*, 113:905–923.

Sickle, J. V. (1989). Dynamics of African ungulate populations with fluctuating density-independent calf survival. *Theoretical Population Biology*. In Press.

Simberloff, D. (1987). The Spotted Owl fracas: Mixing academic, applied and political ecology. *Ecology*, 68:766–772.

Slade, N. A. and Levenson, H. (1982). Estimating population growth rates from stochastic Leslie matrices. *Theoretical Population Biology*, 22:299–308.

Snell, T. W. (1978). Fecundity, developmental time, and population growth rate. *Oecologia (Berl.)*, 32:119–125.

Stearns, S. C. (1976). Life history tactics: a review of the ideas. *Quarterly Review of Biology*, 51:3–47.

Stoto, M. A. (1983). The accuracy of population projections. *Journal of the American Statistical Association*, 78:13–20.

Sykes, Z. M. (1969). Some stochastic versions of the matrix model for population dynamics. *Journal of the American Statistical Association*, 64:111–130.

Templeton, A. R. and Levin, D. A. (1979). Evolutionary consequences of seed pools. *The American Naturalist*, 114:232–249.

Tuljapurkar, S. (1981). Primitivity and convergence to stability. *Journal of Mathematical Biology*, 13:241–246.

Tuljapurkar, S. (1982a). Population dynamics in variable environments. II. Correlated environments, sensitivity analysis and dynamics. *Theoretical Population Biology*, 21:114–140.

Tuljapurkar, S. (1982b). Population dynamics in variable environments. III. Evolutionary dynamics of r-selection. *Theoretical Population Biology*, 21:141–165.

Tuljapurkar, S. (1982c). Why use population entropy? It determines the rate of convergence. *Journal of Mathematical Biology*, 13:325–337.

Tuljapurkar, S. (1984a). Demography in stochastic environments. I. Exact distributions of age-structure. *Journal of Mathematical Biology*, 19:335–350.

Tuljapurkar, S. (1984b). Population dynamics in variable environments IV. Weak ergodicity in the lota equation. *Journal of Mathematical Biology*, 14:221–230.

Tuljapurkar, S. (1985). Population dynamics in variable environments VI. Cyclical environments. *Theoretical Population Biology*, 27:1–17.

Tuljapurkar, S. (1986a). Demographic applications of random matrix products. In Cohen, J. E., Kesten, H., and Newman, C. M., editors, *Random matrices and their applications.* American Mathematical Society.

Tuljapurkar, S. (1986b). Demography in stochastic environments. II. Growth and convergence rates. *Journal of Mathematical Biology*, 24:569–581.

Tuljapurkar, S. (1987). Forecast dynamics in time-series-matrix population models. Paper presented at 1987 annual meeting, Population Association of America.

Tuljapurkar, S. (1989a). Age structure, environmental fluctuations and hermaphroditic sex-allocation. *Heredity.* In Press.

Tuljapurkar, S. (1989b). Delayed reproduction and fitness in variable environments. *Proceedings of the National Academy of Sciences (USA).* In Press.

Tuljapurkar, S. (1989c). An uncertain life: Demography in random environments. *Theoretical Population Biology*, 35:227–294.

Tuljapurkar, S., Deriso, R., and Ginzberg, L. R. (1983). Relative sensitivity of Hudson River Striped Bass to competing sources of mortality and the implications for monitoring programs. Unpublished report to New York Power Authority by ABM, Inc., New York.

Tuljapurkar, S. and Lee, R. D. (1987). Variance decomposition for populations with ARMA vital rates. Unpublished.

Tuljapurkar, S. and Orzack, S. H. (1980). Population dynamics in variable environments. I. Long-run growth rates and extinction. *Theoretical Population Biology*, 18:314–342.

Van Kampen, N. G. (1981). *Stochastic processes in physics and chemistry.* North Holland, Amsterdam.

Wachter, K. (1989). Pre-procreative span, stability and cycling. Graduate Group in Demography, University of California, Berkeley.

Wallace, J. R. (1986). A population model, incorporating environmental variability, of Elk in the Cedar River Watershed, Washington. Master's thesis, University of Washington, Seattle, WA.

Werner, P. A. and Caswell, H. (1977). Population growth rate and age versus stage-distribution models for teasel *(Dipsacus sylvestris Huds.)*. *Ecology*, 58:1103–1111.

Williams, G. C. (1966). *Adaptation and natural selection.* Princeton University Press, Princeton, NJ.

Williams, W. H. and Goodman, M. L. (1971). A simple method for the construction of empirical confidence limits for economic forecasts. *Journal of the American Statistical Association*, 66:752–754.

Ziman, J. M. (1979). *Models of disorder.* Cambridge University Press, Cambridge, UK.

Index

Volume 18

S. A. Levin, Cornell University, Ithaca, NY; **T. G. Hallam, L. J. Gross,** University of Tennessee, Knoxville, TN, USA (Eds.)

Applied Mathematical Ecology

1989. XIV, 489 pp. 114 figs. ISBN 3-540-19465-7

Contents: Introduction. - Resource Management. - Epidemiology: Fundamental Aspects of Epidemiology Case Studies. - Ecotoxicology. - Demography and Population Biology. - Author Index. - Subject Index.

This book builds on the basic framework developed in the earlier volume - "Mathematical Ecology", edited by T. G. Hallam and S. A. Levin, Springer 1986, which lays out the essentials of the subject. In the present book, the applications of mathematical ecology in ecotoxicology, in resource management, and epidemiology are illustrated in detail. The most important features are the case studies, and the interrelatedness of theory and application. There is no comparable text in the literature so far. The reader of the two-volume set will gain an appreciation of the broad scope of mathematical ecology.

Volume 19

J. D. Murray, Oxford University, UK

Mathematical Biology

1989. XIV, 767 pp. 262 figs. ISBN 3-540-19460-6

This textbook gives an in-depth account of the practical use of mathematical modelling in several important and diverse areas in the biomedical sciences.
The emphasis is on what is required to solve the real biological problem. The subject matter is drawn, for example, from population biology, reaction kinetics, biological oscillators and switches, Belousov-Zhabotinskii reaction, neural models, spread of epidemics.
The aim of the book is to provide a thorough training in practical mathematical biology and to show how exciting and novel mathematical challenges arise from a genuine interdisciplinary involvement with the biosciences. It also aims to show how mathematics can contribute to biology and how physical scientists must get involved.
The book also presents a broad view of the field of theoretical and mathematical biology and is a good starting place from which to start genuine interdisciplinary research.

In preparation

Volume 20

J. E. Cohen, Rockefeller University, New York, NY, USA; **F. Briand,** Gland, Switzerland; **C. M. Newman,** University of Arizona, Tucson, AZ, USA

Community Food Webs

Data and Theory

1989. Approx. 300 pp. 46 figs. ISBN 3-540-51129-6

Springer-Verlag Berlin
Heidelberg New York London
Paris Tokyo Hong Kong